Trusted Platform Module Basics

Trusted Platform Module Basics
Using TPM in Embedded Systems
by Steven Kinney

AMSTERDAM ● BOSTON ● HEIDELBERG ● LONDON
● NEW YORK ● OXFORD ● PARIS ● SAN DIEGO
● SAN FRANCISCO ● SINGAPORE ● SYDNEY ● TOKYO

Newness is an imprint of Elsevier

Newnes

Newnes is an imprint of Elsevier
30 Corporate Drive, Suite 400, Burlington, MA 01803, USA
Linacre House, Jordan Hill, Oxford OX2 8DP, UK

Recognizing the importance of preserving what has been written, Elsevier prints its books on acid-free paper whenever possible.

Library of Congress Cataloging-in-Publication Data
Kinney, Steven.
    Trusted platform module basics : using TPM in embedded systems / by Steven Kinney.
        p.   cm.
    Includes bibliographical references and index.
    ISBN-13: 978-0-7506-7960-2
    ISBN-10: 0-7506-7960-3
    1. Embedded computer systems. 2. Computer security. 3. Data encryption
    (Computer science) I. Title.
    TK7895.E42K56 2006
    005.8—dc22

                                                                    2006018308

British Library Cataloguing-in-Publication Data
A catalogue record for this book is available from the British Library.

ISBN 13: 978-0-7506-7960-2
ISBN: 0-7506-7960-3

For information on all Newnes publications visit our Web site at www.books.elsevier.com

Typeset by Charon Tec Ltd, Chennai, India
www.charontec.com

Working together to grow
libraries in developing countries

www.elsevier.com | www.bookaid.org | www.sabre.org

ELSEVIER   BOOK AID
                International   Sabre Foundation

Contents

Acknowledgments

To accomplish the publication of any book, many people must contribute to the transformation of the manuscript into a final text worthy of printing. This one is no exception. First, I would like to thank my wife, Susan Kinney, for her unwavering support during the time of this book's writing. She continues to inspire me through my life's journeys and I would not be half of the individual that I am today without her by my side. In addition, I would like to acknowledge my fellow members in "the original three" – the engineers who initiated the development effort of the first Atmel trusted platform module (TPM). This infamous group, commonly referred to as "The Pit-Bulls," consisted of Spike, Killer and Fido, also known as Eric Gnoske, Nate Bohlmann and me. We resided within an area called "The Pit-Bull Pen" because most of our co-workers just threw raw meat into the pen and kept their limbs away from any openings.

Another individual I would like to acknowledge is Mark Shaw who proofread the final draft of the manuscript before it was submitted to the publisher. He has been with us for 2 years and has proved himself to be a valuable asset and friend. Mark is one of the brightest and most sincere young engineers with whom I have ever had the pleasure of working. In addition, many thanks go out to Steve Motto, the virtual engineering encyclopedia, and Tom Moulton who deals with TPM customers every day and rarely receives kudos for his worthy efforts. Special thanks go to my former boss Randy Mummert, who reviewed the manuscript and always seems to drag me back into the TPM world.

I am also grateful to Al Weiner, my current boss, who helps me stay focused on hardware design and tries to keep the TPM pack off my heels. A special thank you goes out to Tiffany Gasbarrini, my editor at Elsevier, for supplying the guidance, patience and understanding that helped make this book possible. Also, an additional special thank you to Marilyn E. Rash at Elsevier and Mani Prabakaran (Praba) of CharonTec for all their efforts during this book's production. *Trusted Platform Module Basics* is my first book, and I pushed the boundaries with regard to schedule, but Marilyn and Praba guided me through to the finished product.

Finally, sincere thanks to Kerry Maletsky who exemplifies, in my view, a modern-day Renaissance man. He has given me consistently wise guidance, as the best mentors do; and after spending 22 years in this business, I have found that true mentors are hard to come by. Kerry, thanks for all of your efforts in both managerial and technical support. There surely wouldn't have been an Atmel TPM without your input.

Introduction

We live in a digital world that makes our lives easier, more productive and unfortunately more dangerous with regard to identity theft. There is a very big problem with deployed system security that, because of the "security holes," allows individuals to monitor network access points, to hack into operating systems and to install spyware in our personal computers. These are just some of the security pitfalls that affect almost every digital transaction during our daily interactions as customers, employees and computer users.

Some prime examples of the lack of system security can be found within department stores, gaming devices and financial advisors' computers, to name a few. For instance, a major electronics store implemented a wireless network to record cashier transactions into a central accounting hub. The problem was that this implementation wasn't security-based but functionality-based. The wireless network did not employ any type of encryption, so outside in the parking lot, an individual was accessing their network and recording all the credit card transactions.

Another prime example concerning system security failures is Microsoft's Xbox – a system that allows individuals to play content providers' games. The problem with Xbox is that the system relies on obfuscation – "smoke and mirrors" – that, in theory, hide the encryption key from "snoopers" trying to crack the operating system. Well, this poor security protection was cracked and now the Xbox has a Linux distribution targeted at it. Finally, one of the most dreaded security snafus involved on at a major financial corporation who had his notebook computer stolen with more than 200,000 accounts on board. To add insult to injury, the company used a very weak encryption protocol that will potentially allow the thief to gain access to very sensitive account information.

The point is that the design of truly secure systems is no accident; in truth, the lack of a security mind-set, replaced with functional convenience, can lead to security holes that allow sensitive data to be stolen. Have you counted the number of security patches Microsoft has produced in response to holes within its XP operating system. In addition, most deployed systems use a software-based security model, which, by definition, is open to attacks from within by the use of debuggers, logic analyzers and a host of other measurement tools.

Point in fact, a graphic image of the contents within any computers hard drive will divulge the possible keying material stored on the device. The problem is that cryptographic keying material has a very telling signature by virtue of its Hamming weight – the relationship between logic ones and zero within its binary data. To this end, the keys are obvious when looking at a pixel image of a system's hard drive contents, and the keying material can be "phished out" and used against the host systems' security protection.

The trusted platform module (TPM) was designed to bridge the gap between a purely software-based security realization and the added protection that hardware realization adds to the security of any system. By performing all random scheduling algorithm (RSA) private-key cryptographic operations and protecting the RSA private keys within the hardware boundaries, the TPM protects against snooping for keying material contained within a hard drive. Microsoft is designing an operating system that leverages the TPM and more and more original equipment manufacturers (OEMs) are implementing the TPM within system hardware realizations. Embedded systems can benefit from the deployment of the TPM device within their platforms and can use this device for secure bootstrap routines and protect secure application threads running within a system. In conjunction with a secure software stack, the TPM makes a formidable security platform that is much more resistant to software- and hardware-based attacks.

This book introduces some basic concepts of TPM and the command suite used as a level of security that transcends the software only realization. As with any security realization, good system design is paramount and any lapse in security can compromise an entire system. Key management, secret management and other systems-related security parameters must be protected by using sound cryptographic realizations. Also introduced is the TPM interface and suggestions for proper data-management realizations. In addition, this book provides reference materials to use when further cryptographic explanations, which are outside its scope, are needed.

In summary, *Trusted Platform Module Basics* defines the TPM architecture, command compilation procedures and definitions, and execution of commands. This book does not go into the more complex commands; the basic premise of its content is to provide a solid understanding of command compilation and execution. With this knowledge, readers can expand their knowledge of the entire TPM command suite with regard to Versions 1.1 and 1.2 of the device.

The main thing you should take away from this book is the fact that system security is a very difficult task to manage and that everyone needs to be cognizant of system security design and implementation. Never shortchange on system security – the most prevalent means of breaking into a system's security does not lie in attacking the cryptographic engine, but in attacking the data management within it. If an attacker can gain secret and keying material information, there is no need to go "head to head" with the cryptographic engine.

The hope I have here is that you enjoy this book and go away from it with information that can help you further understand the TPM and its capabilities.

TCG Prerequisites

1.1 The Trusted Computing Group

The Trusted Computing Group (TCG) is an organization that develops and produces open specifications, with regard to security-based solutions deployed on various computing realizations. For example, the TCG publishes a group of related specifications that define secure procedures as they relate to the boot-up, configuration management, and application execution for personal computing platforms. These specifications were originally written in reference to personal computers (PCs), but the group is now extending the security realizations into mobile technology and various other embedded system designs. When accessing the TCG web site (www.trustedcomputinggroup.org), various information, along with the group's charter and press releases, can be viewed and downloaded to further educate oneself concerning the security issues facing modern computing systems.

1.2 The TCG Specification Suite

One of the first documents that should be read is the architectural overview, which provides information regarding the TCG goals and the secure system architecture design. In addition, there are two categories of Trusted Platform Module (TPM) based security specifications, the TCG Main Specification Version 1.1b, and the TCG Main Specification Version 1.2. The TCG Main Specification Version 1.1b has been in release form for several years, with various choices regarding the TPM, the Basic Input/Output System (BIOS), the TCG Software Stack (TSS), and application suites. The TCG has various specifications with regard to the Version 1.1b suite and each addresses various aspects of secure system design. Version 1.2 of the TCG Specification Suite is still being defined at the time of writing this book, and can be described as a superset of the Version 1.1b Specification Suite, with some deprecation. Therefore, the study of TPM Version 1.1b-based realizations can be directly applied to TPM Version 1.2-based specifications in the area of what is called *legacy mode*. Since this book is targeted at embedded system design revolving around TPM inclusion, the subject matter is based on the TCG Version 1.1b Main Specification.

The TCG defines a suite of specifications regarding the Version 1.1b TPM-based design realizations. Each of these documents defines a subset functionality that, when combined, produces overall system security protection based on the TPM hardware device. The various Version 1.1 TCG specifications include the Main Specification, the Software Stack Specification, and the PC Specific Specification. In addition, "C" source header files are contained within the Software Stack Specification Header File document. The main focus of the content of this book revolves around the TCG Main Specification, which defines the architecture and command definition of the Version 1.1b TPM. The other specifications are described in general and their content juxtaposed with the typical security needs of an embedded design.

The TCG Software Stack Specification and the PC Specific Specification are also described, with some of the subject matter migrated to the embedded system idiom. Keeping in mind that even though these specifications are written in reference to personal computing, concerning either the physical design or resource management, the security concepts outlined within these specifications can be applied directly or modified to facilitate security within an embedded system. The PC Specific Specification is described first and translates mainly to the bootstrap functionality that is normally found within embedded systems. Of course, if the embedded system is based on PC hardware architecture, the contents of the PC Specific Specification will directly apply to the PC system BIOS.

1.3   The PC Specific Specification and the Embedded Design

The PC Specific Specification Version 1.1 concerns itself with the boot environment on a PC and these requirements are realized within the Memory Absent (MA) and Memory Present (MP) BIOS drivers. The interesting aspects concerning this specification and the embedded design engineer have to do with initial TPM state and Platform Configuration Register (PCR) extensions, which are functionally described within the PC Specific Specification. The key issue for the embedded design engineer is to extrapolate the functional aspects described in reference to BIOS driver development, and to apply these aspects with regard to the embedded systems' bootstrap functionality. The following paragraphs explain the requirements placed on the bootstrap utility of the initial TPM state and PCR digest extension(s). Note that PCR extensions are not strictly required but are recommended because they apply to embedded configuration management.

The first issue is one of TPM initialization, which has various options that determine the post–power-on state of the TPM. The TPM_StartUp command is normally the first command sent to the TPM after host system power-on reset (POR) or after coming out of system hibernation. The TPM_StartUp command is fully described in Chapter 11 about TPM initialization and the low-level command suite, but the high-level aspects of this command are described here.

There are three command modifiers that can alter the TPM state during transmission of the TPM_StartUp command and are defined as default, state, and disabled. The BIOS or bootstrap utility is responsible in determining the TPM_StartUp command modifier that is appropriate relative to host system state; Chapter 11 describes the modifiers and their use in more detail. In addition to initializing TPM state, the TPM_StartUp command invokes an internal TPM partial self-test, which defines the next command that must be sent to the TPM during host system initialization, the TPM_ContinueSelfTest.

Once the bootstrap has invoked the TPM_StartUp command, the TPM will perform a partial self-test within the TPM and expects that the next command message to be a TPM_ContinueSelfTest. If any other TCG message is sent to the TPM, the TPM_ContinueSelfTest will be implicitly invoked, and the command explicitly sent to the TPM will be responded to with a *TPM_RETRY* error code. This information is pertinent to BIOS or bootstrap development and in most TCG-enabled systems, the operating system (OS) or real-time operating system (RTOS) should expect the TPM to be started up while the self-test continues. Referencing the PC Specific Specification Version 1.1, Section 8.2.2 MA Driver addresses the TPM functionality supported by the MA Driver (see Paragraph 8.2.2.4 MA Driver Functions for the functional details).

The next few paragraphs examine the four functions that detail the level of TPM support provided by the MA Driver as defined within the PC Specific Specification, all of which can be migrated to a bootstrap utility "as is" or as a modified version specifically targeted to the host system.

In addition to the "specified" functionality, the "real-world" requirements are addressed and, as we all know, *real-world* design considerations need a real-world TPM. The Atmel AT97SC3201 should be the TPM of choice when addressing any real-world design considerations, including BIOS Drivers and bootstrap utilities. There are a few added functional requirements that are specific to the Atmel TPM and are outlined for inclusion regarding the embedded bootstrap utility or for custom BIOS development. It is worth mentioning that there are BIOS revisions on the market that support the Atmel TPM, and if your BIOS supplier does not offer a TCG-enabled BIOS, Atmel does offer MA and MP assembly source for inclusion within custom BIOS realizations. With that said, the first MA Driver functionality is the MAInitTPM, which covers the TPM_StartUp and TPM_ContinueSelfTest addressed in the previous paragraphs.

The MAInitTPM Driver Function

What the PC Specific Specification Version 1.1 does say, with regard to the MAInitTPM function, is that the TPM must be started up, but what this specification doesn't outline is the specific realizations that must be included concerning the functional aspects defined by the TPM vendor, in this case Atmel. Also, most embedded design realizations are more than likely to leverage the two-wire System Management Bus (SMBus) communication protocol that Atmel offers in lieu of the standard Intel Low Pin Count (LPC) interface. The flow chart shown in Figure 1.1 outlines the functional flow concerning the MAInitTPM in conjunction with an onboard Atmel AT97SC3201 TPM.

Ignoring all of the PC BIOS issues and concentrating on specific TPM status, configuration, and command requirements, the following must happen to satisfy the specified MAInitTPM functionality as it relates to the Atmel AT97SC3101 TPM. First, the TPM status register must be checked to verify that the TPM is able to accept configuration modification and TCG command messages. This is accomplished by reading the LPC input/output (I/O) address baseAddress1 + 1, which equates to the hexadecimal value 4F. If the contents of the status register come back as hexadecimal value BC, then the

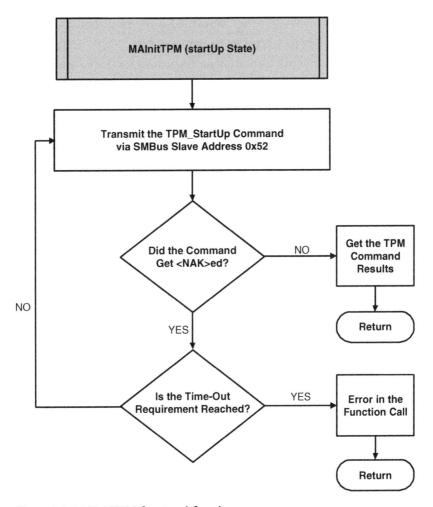

Figure 1.1 MAInitTPM functional flow diagram

TPM is not ready to receive any type of command and the BIOS or bootstrap utility must continue to poll the TPM status register. Note that this procedure is confined to the LPC protocol interface, the SMBus protocol interface; slave address hexadecimal value 52 will simply Not Acknowledge (NAK) the master if the TPM is not ready to respond to requests. Specific LPC and SMBus concerns are addressed in Chapter 3 about hardware considerations and TPM communication.

The SMBus is the specific TPM communication protocol referenced throughout this book because the LPC bus is well defined, both by the TCG and Atmel-specific documentation. Considering that it is tailored to embedded design, without a doubt, the SMBus will be the main interest of the book's readers. Again, if you're getting "bummed out" over the lack of specific LPC bus discussion, refer to Chapter 3 for more detail. If you're still not satisfied, refer the multitude of documentation regarding the LPC bus that is available; start with your vendor regarding the TPM.

With that said, when the MA Driver is accessing the TPM via specific functional calls, the metric is determining that the device is present and available, as provided by the SMBus protocol itself. For example, if the MA Driver is sending a TPM_StartUp command via the SMBus, the TPM will ACKnowledge (ACK) the command and execute, if not, the TPM will simply NAK the command. If the developer "wraps" this logic in a time-sensitive function, the call can be aborted and the TPM "issue" can be addressed. Hence, the TPM SMBus protocol has a very simple interface, with regard to the TPM access, an ACK, regarding the TPM slave address, means the TPM is available and can accept further command-specific data. On the other hand, if the TPM NAKs the master's attempt to communicate with the TPM, the master can "poll" the TPM by sending repeated TPM slave address inquiries until the TPM acknowledges the request or the function times out.

The TPM communication check can be integrated within the main requirement of the MAInitTPM main functional specification to initialize the TPM by sending a TPM_StartUp command. If the command is ACKed, concerning the SMBus protocol, the TPM has accepted the command and is therefore present and available. On the other hand, if the MAInitTPM call produces a time-out due to multiple NAKs, this will alert the host system and design engineers that there is a communication issue regarding the TPM. Again, the main goal of the MAInitTPM is to initialize the TPM and test the communication protocol concerning this resource.

The MAHashAllExtendTPM Driver Function

This next MA Driver function call is in regard to host system configuration in which a PCR, indexed by the passed argument, can be extended and updated. The functional description regarding this driver's calls implementation is defined in Figure 1.2.

Simply put, the entity making the call, in the case specified by the TCG, the PC BIOS will supply parameters regarding the start of the location to hash-extend, the length of the data to hash-extend, and the PCR index that the resulting digest is to be extended to. A side note, this verbiage is intended to acquaint the reader with the specified functionality regarding the PC BIOS defined by the TCG. With this knowledge, the embedded design engineer can

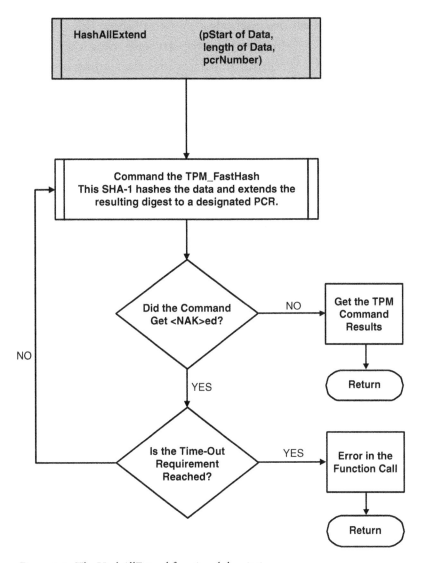

Figure 1.2 The HashAllExtend functional description

modify the specific implementation to suit their needs without architecting a complete solution from the ground up. For more specific information regarding the SHA-1 and PCR extend capabilities of the TPM, see the chapters devoted to these command-specific requirements.

The MAPhysicalPresenceTPM Driver Function

The final function associated with the MA Driver concerns the subject of Physical Presence. This is a very ambiguous subject since most TPM vendors have implemented various methods for accomplishing this goal, which results in the protection of Intellectual Property and thus the Non-Disclosure Agreement (NDA). What this translates to is the signing of an NDA before the TPM vendor will disclose their individual Specific Command implementations for which TSC_PhysicalPresence is part. Nonetheless, the MA Driver does support a call that will allow the establishment of TPM Physical Presence as shown in Figure 1.3.

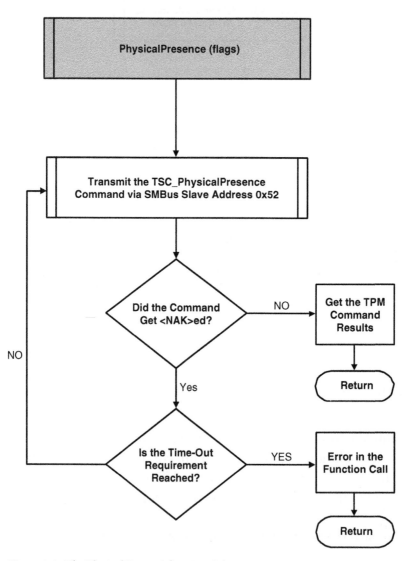

Figure 1.3 The Physical Presence functional description

To get more information about the specific functionality and variations regarding this Vendor Specific Command, as well as the full description concerning the suite of specific commands, contact your Atmel sales representative. In addition, the TPM specifics concerning the concept of TPM Physical Presence are discussed in Chapter 11 about TPM initialization.

The Memory Present TPM Command Transmit Function

The Memory Present Driver, the driver associated with the MA Driver, has four function call definitions with only one of these being of interest regarding embedded systems. For full knowledge of the MP Driver functional support and implementation details, refer to the PC Specific Specification itself. The one important function defined within the Memory Present Driver chapter about the PC Specific Specification involves the transmission of TPM commands. This functionality loses value, especially considering the design of embedded bootstrap code, when the Trusted Device Driver Layer (TDDL) is present within the system (see Chapter 22 on this driver's functional detail). The reason that this may be so involves the case of the embedded system, where the bootstrap functionality is "lean-and-mean" with as little overhead as possible. Hence these types of systems may have a MA and TDDL support or forgo the MP, strictly leveraging the interface to the TPM defined by the MA and TDDL. If your system's resources are available on power-up and do not use a phased boot procedure, I would recommend using the MA and TDDL directly without the added overhead (increased code size) that the MP Driver would add with little or no system benefit. However, the PC Specific Specification does add valuable insight to TPM and host system security that would aid in the design of embedded bootstrap utilities.

1.4 The TSS Specification Version 1.1

The final specification with regard to embedded system design and the TPM involves the TSS. This specification is important to embedded design engineers since it defines the TDDL, a driver that will undoubtedly be leveraged within the embedded environment. In addition, the TSS gives insight into the construction of a security software stack, which adds an abstraction layer to the TPM command interface. When integrating the TPM into any system design, not only do the specific TPM commands have to be managed, but also cryptographic support must be in place to aid in the construction of the commands. Some TPM vendors design in "helper" commands that augment the TPM command suite in the form of vendor-specific commands. Before any attempt is made to design an "Embedded" TSS, your TPM vendor should be contacted concerning the vendor-specific commands that might make this tasks easier, both from a time and a resource metric.

In summary, the TCG is an invaluable resource regarding research when investigating embedded design solutions involving the TPM. The group's web site has numerous specifications, articles, and white papers about secure system architecture. TCG membership might be considered if your company is involved in product design for platform security, especially if the security considerations involve PDAs or mobile phone technology. There are TCG working groups specifically addressing the security issues that involve the above-mentioned technologies. The bonus of joining a group is that one or more company representatives can take part in discussions about specific TCG specifications that may affect the company's product security model.

2

Cryptographic Basics

To interface with the Trusted Platform Module (TPM) and produce meaningful command content, you must have an application level of understanding with regard to cryptography. This implies that you are able to create encryption and signature blobs, as well as decrypting and verifying these cryptographic entities. The most prevalent cryptographic system employed by any vendor's TPM is the RSA. This cryptographic system is employed through input and output command messaging as well as the internal protection profile. Other cryptographic systems that vary in support, in regard to individual TPM vendor realizations, involve the Data Encryption Standard (DES), the Triple Data Encryption Standard (3DES), and the Advanced Encryption Standard (AES) to name a few.

The first thing that must be realized is that each of the systems can be placed in two separate categories: symmetric and asymmetric key-based cryptography. This discussion about cryptography focuses on the 30,000-foot view, or application-specific approach, as opposed to the fine implementation details. For a more thorough discussion with an expert's explanation of cryptographic systems, see the reference at end of the chapter.

The first section describes the two distinct categories of cryptographic keying material: symmetric and asymmetric. After this we discuss the application with regard to the best cryptosystem for the task at hand, whether that is file encryption, signature generation, or keying material protection.

2.1 The Symmetric and Asymmetric Keys

Symmetric keys refer to the use of a single cryptographic key for both encryption and decryption. Examples of this type of cryptosystem are DES, 3DES, and AES. This simply means that the same key will be used, from the perspective of the application and not internal to the cryptographic algorithm, to encrypt data as is used to decrypt data. This cryptographic property makes the handling, or key management, somewhat difficult, considering that the knowledge of this key will allow individuals other than the owner of the key to decrypt the encrypted message. For example, Fred is somewhat daft and he encrypts his diary with a 3DES symmetric, key, but he wants his brother to be able to read his diary because they are very close. The only secure means that Fred has, at this juncture in our discussion, that would allow his brother to read the diary is to physically meet him and transfer the 3DES key to him personally. If Fred sent the key via email, anyone could read the message, either by intercepting it or by looking over his brother's shoulder; the point is, it's not secure. But this type of cryptosystem is very fast with regard to the encryption of large amounts of data and Fred is looking for a solution that will allow him to keep the symmetric key and add a level of security to protect its contents.

Enter the asymmetric key pair used within a cryptosystem; it has the notion of both a private and public key, hence the term "key pair". The most popular asymmetric key-based cryptosystem is RSA and the TPM is no exception regarding the use of this popular cryptographic algorithm. RSA uses two keys: a private key to decrypt or sign data and a public key to encrypt or verify signatures. The RSA private key is to remain secret and only the owner of this key should know its value. The public key, by virtue of it name, is known to all and can be freely distributed. Therefore, Fred can get a copy of his brother's RSA public key and encrypt his diary with this key. Fred can rest assured that his diary will be seen only by his brother and no one else, since the encrypted message can only be decrypted by the private key associated with the public key Fred used to encrypt the diary. Fred can even have his brother sign an email using his private key, acknowledging that he decrypted the dairy, and Fred can determine that this email is valid by verifying the RSA signature with his brother's public key.

The main point is that the asymmetric key pair is much more easily manageable and facilitates the use of a public key, and in doing so, protects the private key from being discovered. In our little example, Fred is daft and he used an RSA cryptosystem to encrypt a large amount of data, which works, but is not the most efficient means of file encryption. What Fred could have done is to encrypt the diary using a symmetric key leveraging the 3DES algorithm. Then Fred could send the encrypted file to his brother and use his brother's RSA public key to encrypt the 3DES key and send this information as well.

Fred's brother would get the encrypted diary and the 3DES key encrypted with his RSA public key. He would then decrypt the 3DES key with his RSA private key and use the 3DES key to decrypt the diary. The fact is, RSA is about a thousand times slower than 3DES, but 3DES does not lend its keying material to easy management schemes. So the TPM leverages both strengths and uses RSA to manage RSA private keys and any other entities that must be protected; symmetric keys, secrets, and other data are considered private. In fact, no RSA private key is allowed to migrate from the TPM in clear text form or a form that would allow someone to ascertain its value just by looking at the data.

With this level of key management, the host system can leverage the TPM in order to protect a symmetric key and later ask the TPM to supply the symmetric key so that the system can decrypt a file using 3DES or whatever symmetric algorithm is chosen. Basically, the user can bind and encrypt the symmetric key using one of the TPM RSA public keys and later command the TPM to return the symmetric key by internally decrypting the encrypted blob, thereby producing the symmetric key value.

This type of keying material and data protection is used throughout the TPM Command Suite. For example, when loading RSA keying material into the TPM, the command needs to use a parent key – an RSA key pair whose public key will be used to encrypt private data associated with the key about to be loaded. One of the parameters that will be encrypted is the child key; the key about to be loaded into the TPM, private key data and this data will be decrypted internally within the TPM. Once the private key is decrypted, the child key can be loaded into the TPM and the private key will only be known to the TPM, if the TPM generated the RSA key and wrapped this key internally using the parent's public key (see the discussion concerning the TPM_CreateWrapKey command). The bottom line is that the TPM leverages the RSA public key to a private key relationship to create, load, and migrate keying material to and from the TPM in a very secure fashion.

This discussion is not very rigorous, but a generic view concerning the use of a cryptographic system is sometimes the best place to start. The basic cryptographic functionality needed by applications communicating with the TPM is RSA encrypt and signature verification or public key decryption. These cryptographic functions make use of the RSA public key; remember, the private key is either physically or logically "locked" inside the TPM. Therefore the most widely used RSA cryptographic function, for applications communicating with the TPM via command messaging is the RSA public key encryption.

With this, I would like to give some examples that leverage this concept at a level of abstraction above the TPM command itself, focusing on the cryptographic procedure when building TPM commands.

2.2   Using RSA to Encrypt Private Information

This concept is leveraged through TPM command compilation and is used to protect private keying material, symmetric keying material, and sensitive data during command message transmission between the TPM and the secure stack. With that said, let's look at an example that uses an RSA public key to encrypt a 20-byte secret that the TPM will later decrypt internally. Figure 2.1 describes this procedure using abstract functional blocks that

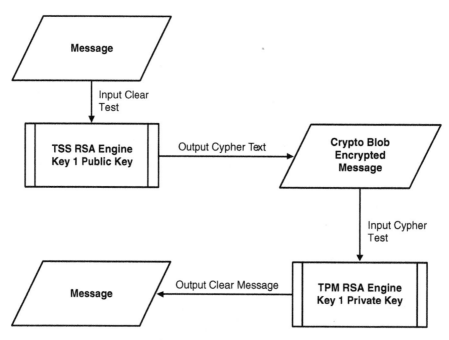

Figure 2.1 RSA encryption and internal TPM RSA decryption

show the security stack encrypting the secret, and the TPM internal RSA engine decrypting the secret.

In the example, the secret is encrypted by using the public key data provided by the TPM, and the secret is decrypted within the TPM after the Trusted Computing Group (TCG) command has been sent by using the associated private key. The TPM will never allow the internal private keying material to leave its boundaries in a "clear" state, a form that will allow immediate use of this private key.

Any private key, existing within the constraints of a physical TPM shielded location must be protected, via RSA encryption, prior to allowing this data to migrate from the TPM. This brings us to the concept of "wrapping" RSA private keys, which is used during TPM command execution for creating, loading, and migrating of RSA keying material to and from the TPM device. Figure 2.2 defines the construction of an RSA key that will be digested by a TPM in regard to public and private data that is protected by a parent key used as an RSA encryption vehicle.

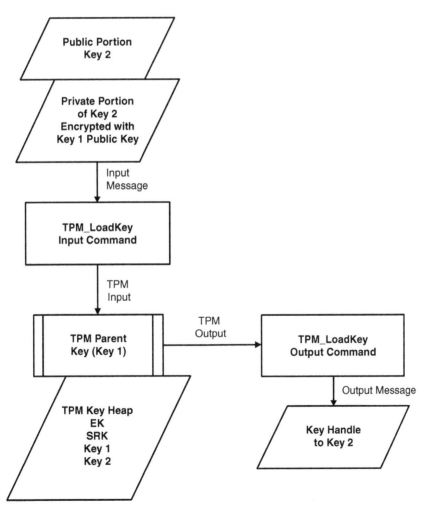

Figure 2.2 RSA encryption used to protect private key data

The preceding figure shows that the keying material exists in two distinct forms, the public data that is in the clear and the private data that is encrypted with a parent's public key and takes the form of an RSA encrypted blob. This form, relative to keying material, is the standard form representing the payload used in loading a key into the TPM. In addition, if a RSA key pair is generated within the TPM, the output of this command – supplying the generated key to the caller – is also in this standard form and can be used directly by the command that loads keying material into the TPM.

One other issue that should be generically described concerns key migration, either migrating from one TPM to another or migrating a TPM key to a third party who will archive this keying material. This concept is no different than the discussion concerning the loading of keying material into a given TPM that has a specific RSA key internally loaded to facilitate parent key functionality. The parent key functionality is simply defined as protecting the private data of an RSA key pair by encrypting the private data. This means that the parent key must be resident within the given TPM to facilitate the decryption of the private data, which will allow the child key to be loaded into the TPM. *Child key* refers to the key that is about to be loaded into the TPM.

Migration is simply the function of "un-wrapping" the RSA key's private data and "re-wrapping" this data with another parent key. Note that the unwrapping function applies to the parent's private key that exists within the TPM that the key is to be migrated from and the wrapping function applies to the new parent's public key whose private key exists within the TPM that the key is to be migrated to. The only difference between migrating to another TPM and a third party is the form of the structure containing the private data; we will talk about these differences in Chapter 21 about key migration.

The important aspect to take away from this discussion is the fact that migration simply involves the redefinition of the parent key, which protects the private data concerning the key to migrate. Figure 2.3 shows this concept with regard to the migration of an RSA key pair from one TPM to another. This concept will help the reader navigate through the command details about specific implementation of migration commands using the TPM.

2.3   Using RSA to Sign and Verify Signatures

In addition to encryption support, the TPM also supports the signing of data. A signature can be added to the data itself or to the SHA-1 hash of the

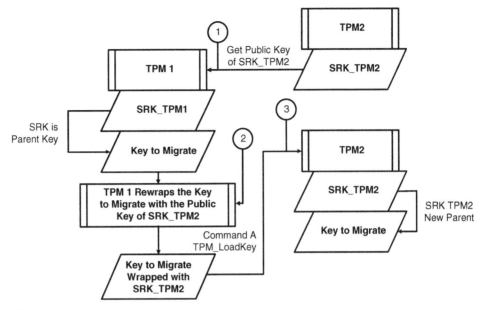

① Get the public key in regard to the new parent SRK_TPM2.
② Rewrap the key to migrate with the new parent, SRK_TPM2.
③ Load the newly wrapped key into TPM2 via a TPM_LoadKey command.

Figure 2.3 Migrating a RSA key from one TPM to another

data. A signature on the SHA-1 hash digest allows the message to be of any length, since the signature is performed on a consistent length of 20 bytes. On the other hand, a signature on the message directly limits the length of the message because the maximum length of the message is bound to the size of the RSA public key. Further limiting data length that can be signed, the encoding scheme will subtract from the maximum area that can be signed. For example, if you are trying to sign a message using a 2048-bit RSA key whose public key size is 256 bytes, the maximum length is 256 bytes minus the encoding parameters, in this case 11 bytes. Therefore the maximum length of the data message that can be signed directly is 254 bytes. The moral of the story; if your message lengths vary to include messages larger than 254 bytes, sign the SHA-1 digest of the messages.

The SHA-1 digest is a one-to-one representation of the message and the signature is attesting to the message's uniqueness and the signature of the message's digest is equivalent. An RSA signature is equivalent to encrypting with a private key; remember that we encrypted data with the public key.

The reason that the signature is produced with the private key concerns the procedure used to verify the signature, a public key decryption. So if Fred sends me a signed email, I can be confident that Fred sent me the message and the content represents Fred's intent by verifying and decrypting the signature

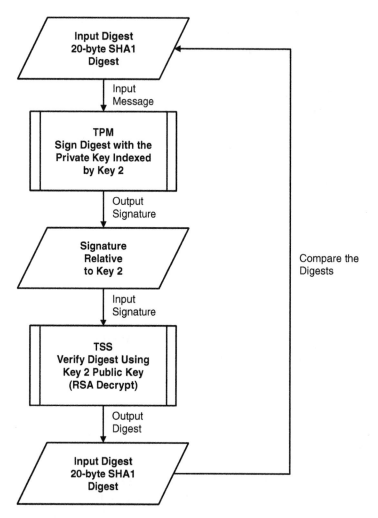

Figure 2.4 RSA signature generation and verification

with Fred's public key. Figure 2.4 shows the concept of signature generation and verification with respect to the specified TPM Command Suite.

Note that the TPM only deals with private keys with regard to commands that support signature operation. This implies that the only "TCG-specified" command is that of signature generation; the verification of signatures is defined within the TSS Specification. This makes sense since the verification of signatures is indeed a public key RSA operation and the TPM is primarily concerned with private key operations. An additional example of this concerns the binding or encrypting of data, another RSA public key operation that is supported by the TSS stack, not the TPM as specified by the TCG. With that said, you should contact your TPM vendor in regard to "vendor-specific commands" that may help the embedded design engineer with RSA operations normally defined at the TSS layer. Most of the specifics are covered by an Non-Disclosure Agreement and the vendor will provide all the information, including vendor-specific commands, after one signs this document.

In summary, the RSA encrypt and signature verification is a public key operation that is normally performed outside of the TPM as defined by the TSS Specification. The RSA decrypt and signature operations are private key operations performed within the TPM, since the private key can never leave the TPM in the clear. All specific TPM commands concerning the execution of specific RSA operations will be addressed within the context of this book. If you are looking for a more detailed cryptographic discussion, please refer to the Schneier book.

Reference

Bruce Schneier (1996). *Applied Cryptography, 2nd Edition*, New York: John Wiley & Sons.

Overview of the TPM Architecture

It would be fairly difficult to discuss the fine details concerning Trusted Platform Module (TPM) functionality without first understanding its architectural design. This chapter covers a generic TPM architecture, as defined by the Trusted Computing Group (TCG), and discusses each element in detail. All TPM realizations must conform to certain specifications with the goal of producing an element of trust with regard to this device. To be more specific, would you blindly use a system to handle your bank accounts or credit card transactions without some element of trust concerning the means by which your personal information is being protected? In addition, you would probably demand proof, in the form of some kind of compliance report, that the device is indeed trustworthy. This book is not going to describe the procedure that the TCG uses to test the compliance of any given TPM relative to the design specifications prescribed. On the other hand, it really would be useful to understand the protection and functional architectural principles leveraged while designing the TPM.

3.1 The TPM CPU or Microcontroller

All TPM realizations must have the ability to perform sequential operations, therefore the TPM architecture must contain a central processing unit (CPU) or a microcontroller. Different TPMs contain various central processing realizations and it is totally the preference of each vendor to produce this type of device. It is safe to say that the key requirements needed include a CPU, non-volatile memory, volatile memory, and input/output (I/O) communication support. These are the bare minimum components; others could include co-processors that perform specific tasks related to an RSA—the main cryptographic engine used by the TPM; it was described first by Rivest, Shamir, and Adleman in 1977, resulting in its acronym-only name. Such components are considered to be foundational, as opposed to secure. A security layer, either in the form of hardware or firmware, wraps the foundational elements with the single purpose of guarding access to the system's functions or data. Figure 3.1 shows a generic TPM-system engine, designed to accommodate the requirements set forth by the TCG.

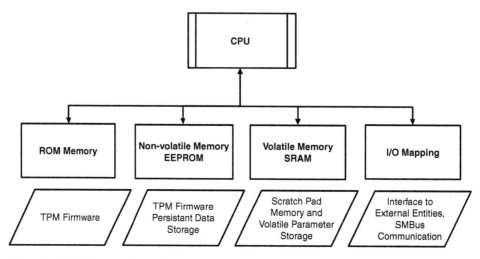

Figure 3.1 TPM computing engine

The figure depicts a central processor interfacing with the typical memory blocks normally associated with this type of design. In addition, the system has supports I/O interaction, external tamper detection, and RSA computations. Is in turn describes a generic flavor concerning the design realization that might be produced by a TPM manufacturer, but the conceptual elements reflect a true system design. Vendors are free to design in "vendor-specific" blocks that augment the TPM system's functional and security aspects in the form of acceleration or added security layers.

3.2 Asymmetric Functional Block Requirements

There are certain requirements that must be realized within the TPM regarding asymmetric functionality, in other words RSA-specific functional support. In this definition, I expand the boundary to include discussion of hash algorithms, since these functional blocks are closely related. The TPM must be able to produce a number of cryptographic entities and functions regarding the RSA. One basic element is concerned with the ability to generate a random sequence of values, unless of course you would like to use predictable values within your secure device. I can hear my marketing manager saying now, "Can we patent that?" The TPM generates random numbers by referring to a hardware block that produces random noise, which can be sampled to acquire a variable-length random seed. This random seed can then be supplied to a pseudo-number generation algorithm, which produces a unique random number sequence. It is worth mentioning that the pseudo-random number generator is just the same as the pseudo-algorithm that produces the same data when given a constant seed.

The moral of this story is the randomness of any value produced by a firmware-based function, purported to be random, and is based on the quality of the seed supplied to this algorithm. Therefore, excruciating effort is focused on the goal of designing a hardware block that produce truly random data and is an integral part of any cryptographic system. Before we discuss this further, it is essential to mention that the TPM gives access to its random number-generation block via TCG command(s). Therefore all embedded system types can leverage the TPM and not worry about firmware-based random number generation.

Leveraging random number generation, the next topic of discussion, concerns the generation of RSA key pairs. This requirement defines the bulk of the co-processor design, which gives the TPM the ability to perform asymmetric cryptographie functions. One of the functions involves the generation of an RSA key pair; "hey you can't dance if you don't have any feet". The co-processor generates RSA key pairs and gives them to the TPM CPU to access this information, which allows the TPM to securely store this material. The procedure by which RSA keys are created is beyond the scope of this book, but be aware that this involves checking randomness, prime number validation, and other RSA-specific requirements. The co-processor is also tasked with the ability to encrypt and sign data using externally or internally generated RSA keys. For example, the TPM CPU can present the co-processor with an RSA private key along with some data and command the encryption or signature of that data. The co-processor will do the requested function and supply the CPU with the resultant blob—crypto-speak for cipher text—and the CPU can do whatever is necessary with regard to this datum. Figure 3.2 shows a simple CPU to co-processor relationship within the context of the TPM.

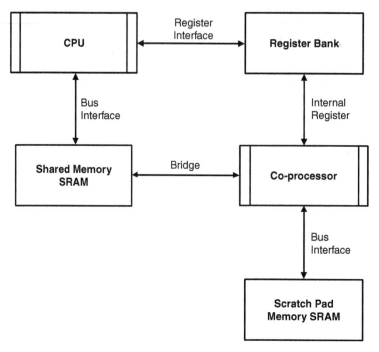

Figure 3.2 TPM CPU co-processor relationship

As promised, there are a few extra advantages concerning the TPM crypto-graphic support functional block and those will be discussed in other chapters. These functions are not cryptographic per se, but involve func-tionality pertaining to hashing data producing a result defined as a digest. The two most relevant algorithms involve the Key-Hashed Message Authen-tication (HMAC) code and Secure Hash Standard SHA-1 hash functions. Simply put, these functions input a message of varying length and produce a 20-byte "one-way" digest. The term *one-way* infers that the message used to produce the hash digest cannot be reproduced by using some reversal process or function. Hash digests are used to uniquely identify a large block of data without divulging the specific content. For example, if you hash a document, producing a 20-byte digest representing the specific content and later re-hash that document, producing a different 20-byte digest, you would know that the document has been altered. The TPM supports both the SHA-1 and HMAC hashing algorithms, which are leveraged quite often during command-execution and authorization protocols. The SHA-1 is a simple "message-in–digest-out" algorithm and the HMAC is a SHA-1 algorithm with an additional element, a 20-byte key.

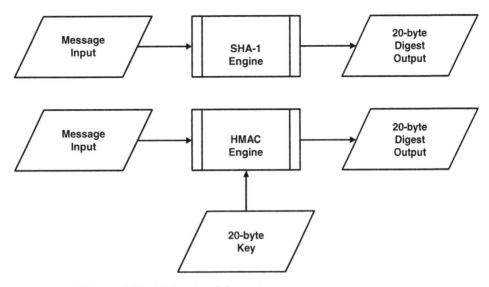

Figure 3.3 SHA-1 and HMAC functional description

Figure 3.3 shows overall functionality concerning these two algorithms. This is regarding the generic view of TPM asymmetric services. Specific command discussion will shed more light on the specifics regarding the asymmetric command suite. In addition, different TPM vendors design various additional functionalities for cryptographic support, hence it would be wise to contact your TPM supplier.

3.3 TPM Memory Blocks

First, let's discuss the non-volatile memory, which is used to store persistent data and state, residing within the TPM. The TPM needs the ability to store keying material, authorization-based data, and internal state; all must survive a power cycle event. Hence the non-volatile memory block; for those of you scratching your head, this type of memory is usually associated with electrically erasable, programmable, read-only memory (EEPROM) and FLASH memory devices. If you are still scratching your head, all right PC guys, this type of memory is indicative of the hard drive within your personal computer. The TPM sets aside various partitions that store individual data classifications based on the type of data being stored. For example, there is a key heap for RSA key pairs, a secret storage area, a TPM state storage area, and other areas that store various data entities. The point is that the TPM must keep certain information within its boundaries away from (!do-gooders) who seek such information for malicious purposes; it must not be cleared on cycling of power. An example concerns RSA private keys, which can never leave the TPM in the clear per specification and must be retained during power cycles.

In addition to non-volatile memory, there is a block of memory that defines volatile or random access memory (RAM). All of you are aware that computations within any given computer-based system must have an area in which to perform operations on data that is fast and convenient; the TPM is no exception. Examples of this type of memory usage involve the calculation of the authorization digest—a digest that validates command execution based on input data and knowledge of a secret. Performing this type of calculation would be slow with regard to using non-volatile memory, not to mention endurance issues associated with this type of memory. The bottom line is that the TPM uses both non-volatile memory for persistent storage and volatile memory for calculation-based operations.

3.4   Platform Configuration Registers

One subject worth mentioning here is the idea of a Platform Configuration Register (PCR). The main purpose of these registers – there are 16 in total – is to attest to the expected system configuration or alteration thereof. The registers are 20 bytes wide and contain a hash digest representing the "state" of a particular hardware or software entity. Registers are stored within non-volatile memory and can be leveraged via various TCG command(s) to set and validate their contents with the goal of testing system-level alteration. PCR metrics in conjunction with the asymmetric co-processor build on a significant security base to ensure that the deployed embedded system has the same configuration as when it was delivered. The subject of PCRs is so important regarding system alteration detection that it is worth mentioning within the context of the TPM system overview. Figure 3.4 shows a basic PCR configuration scheme that includes the various components that come into play with regard to host system configuration management.

PCR handling will be discussed in detail in later chapters. The goal of this section is to make the reader aware of the concept concerning PCR usage and to keep in mind that the RSA cryptographic functionality is not the only host system security game in town.

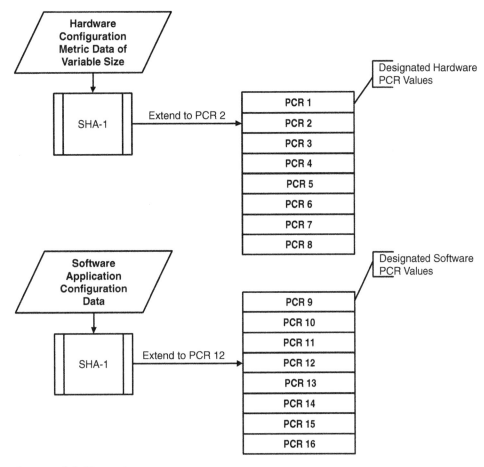

Figure 3.4 PCR overview

3.5   Hardware Power Management and Tamper Circuitry

Power management is a very big concern with regard to the TPM and its security profile. The ability to react to cases in which the TPM loses main and battery power planes is a cause of concern regarding the TPM security profile. In addition, some host systems supply only main power, Vcc, to the TPM, which must be determined by the TPM so that the proper power management profiles can be applied based on the TPM system deployment. For example, if the TPM is notified that battery power has been applied to its Vbb pin and there is a loss of Vbb power, the TPM will flush and clear the TPM of all keying material, except the Endorsement Key (EK) and ownership state.

Now don't go "wigging-out" on me; I said *if the TPM is notified* that there is battery power available. Any system that supplies only a Vcc power plane to the TPM would simply make sure that the register used to notify the TPM of this state indicates that battery power has not been connected. Since no battery power is the default state, your worries about issues during development are nothing but that, worries. If you do have battery power and want to notify the TPM of this state, please see Chapter 11, TPM Initialization.

Another concern is that of physical attacks posed against the TPM in an effort to gain secret information contained in the device. This is a serious threat that can manifest in various forms, including current signature, a subcategory concerning Differential Power Analysis (DPA), physical probing, temperature profile attacks, and clock/voltage manipulation. The point is that the TPM must have the ability to thwart attacks by means that protect information stored within its boundaries. Most TPMs have what is called tamper circuitry – that is, analog circuitry that detects the malicious physical attempts to gain access to sensitive data. Again, these analog protection schemes are vendor-specific and privileged concerning disclosure within the context of this book. Just be aware that protection schemes are present and that the TPM architectural overview – see the next few pages – depicts this function as a black box.

3.6   The TPM, System-on-a-Chip

Finally, the entire TPM is laid out, combining the various design blocks into a full system overview. The TPM is called a *system-on-a-chip* or computer system running embedded code performing a specific task to provide security-based functionality. With this said, keep in mind that the TPM is not passive; it will perform TCG commands and protect itself if necessary. The TPM is autonomous concerning its system functionality; it relies on its own resources to perform defined operations and monitors some external stimuli to defend its secrets. This is important with regard to host system application or secure stack development – something I'm sure you will be doing given the fact that you are reading this book. There are some cases when you do not want to take a pure empirical approach with regard to application development for TPM-based commands. The TPM will limit the command suite and evoke vendor-specific penalties in response to failed command authorization attempts. Much more information will be given later regarding the approach that should be taken when developing code that invokes TPM command messaging. For now, let's look at the overall TPM system internal generic design shown in Figure 3.5.

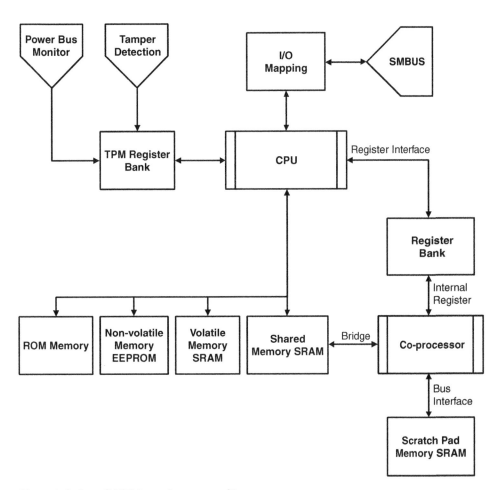

Figure 3.5 Overall TPM generic system architecture

The main concept that needs reinforcement is that the TPM is a system in and of itself, and interfacing with this system requires knowledge of its capabilities and protection profile. The TPM retains secret information along with RSA private keying material that can be leveraged, but not disclosed for any reason. TPM vendors can augment the design to include various "enhancements" to its security profile or accelerate command execution. A vendor representative can disclose TPM vendor-specific functionality in response to the requestor signing a non-disclosure statement. This is fairly typical within the TPM world since the nature of these modifications involves proprietary solutions.

With this said, discussion of some topics is not permitted within the context of this book, but one may contact Atmel to receive additional vendor-specific information about its TPM realization. If there is a unique vendor-specific solution that will help in the development of an embedded system, I will note this exception to alert the reader to investigate by contacting the TPM vendor. The next chapter adds to this discussion by defining the "root-of-trust" designed into the TPM and in what form the trust manifests itself.

Root-of-Trust—the TPM Endorsement Key

Now comes the time to talk about trust – who do you trust? This is a very big question, and the answer depends on the person or entity along with what is being entrusted. For example, Joe Crook, who just got out of the slammer, has opened Joe's "No Questions Asked" used car lot and when you ask the question "Are these cars hot", he responds "Trust me". To put a fine point on the matter, your boss asks you to develop a trusted system leveraging the ACME Trusted Platform Module (TPM) – "Hey, a TPM is a TPM, right". Wrong!

Let me ask the question, "How do you know that the TPM you are about to design in is yea and verily a Trusted Computing Group (TCG) certified device?" Some marketing guy saying, "Trust me". Yea right. This brings up another good question, "How do you know that the TPM sitting on your desk or designed into your system is what you think it is, a secure TPM that was manufactured by the Atmel Corporation?" Better yet, "How do I distinguish TPM A from TPM B, especially if one would like to identify the TPM based on cryptographic data, which is the primary means of identification regarding one.

4.1 Root-of-Trust

What is being defined here is a concept called "root-of-trust". Anything or anyone deemed "trustworthy" has some undeniable characteristic or qualification that instills this sense of trustworthiness. I can accept that something is true if someone who, or something that, professes this truth appears, from my perspective, to be reliable. For example, in a court of law my lawyer is an extension or representative concerning my interests. The court trusts that the lawyer is acting on my behalf because of his license to practice law and the client-to-lawyer relationship that has been established by the legal system. The point is that there is a well-defined proof, the license to practice law, and a well-defined procedural process, in this case the legal system. OK, that might be debatable. What happens when I want to sign a legal document at a remote location relative to the person requiring my signature? I simply go to my local bank and sign the document in the presence of a notary and all is good. "What makes a notary so special?" This individual has been deemed reliable in validating the identity of the person who is to supply the signature and witnessing the signature of said person. The notary has two characteristics that make this procedure work, root-of-trust – the legal system has set up a procedure for this person to become a notary – and the procedure itself – identity check and verify signature.

4.2   The Endorsement Key

Well, the TPM also has a root-of-trust that is defined by the Endorsement Key (EK) pair, the unique RSA key found within all TPM devices. Now the EK alone does not prove that the TPM device is trustworthy from the standpoint that this is a valid Atmel TPM; it is simply an RSA key pair with a public and a private portion. The EK does, on the other hand, identify a unique TPM – I can read the EK public key and use this information to select a single TPM from a pile of any number of TPMs regardless of manufacturer. This is an important step in determining the trustworthiness of the TPM on my desk; I can identify this TPM by use of a cryptographic means. The only thing missing is a method to prove that the TPM with an EK public key of X has been manufactured by Atmel. This proof involves a third party, someone other than the TPM on my desk or myself. That third party is the Atmel Corporation.

When a TPM is manufactured, an EK pair is created; there are various methods with regard to when the EK is created, but by having Atmel create the EK, it is much simpler to explain the EK root-of-trust relationship. Most reliable TPM vendors offer an EK-generation service. Another fact concerning the EK is that, once it has been created, it can never be replaced or removed from the TPM. In other words, the EK is a nonmigratable RSA key pair. This makes sense; if you can remove or replace the EK, you have just eliminated the unique identifier from the TPM. When Atmel creates the EK, the public key is managed and the private key never leaves the TPM; the recorded

EK public key is in a secure form with no other identifier. This means that I can give Atmel a call and say, "I have a TPM on my desk with EK public key (X), is this one of yours?" Atmel compares the EK public key with the known EK database and says, "Why yes it is." This scenario has completed the root-of-trust – I can uniquely identify the device in question and I can ask a corporation that I trust to profess that this is indeed a device manufactured by that corporation.

4.3   X509 Certificate

The proceeding is a very simplistic explanation, but the information gathered and the ability of Atmel to identify the device is accurate. Regardless, there must be a better method than picking up the phone and dialing up the TPM manufacturer to ascertain the origin of the device in question; indeed there is. Introducing the concept of a certificate, an X509 Certificate has to be exact. The X509 Certificate alone, like everything else, is not sufficient to warrant trust. Unless Joe Crook's No Questions Asked kind of certificate appeals to you. That's right, the issue of trust also pertains to the origin of the certificate and this marks a good place to introduce the Certification Authority or CA for short.

The CA is an entity, not unlike a notary in the previous section that produces X509 Certificates and conveys the element of trust between parties who do not trust each other. For example, you get a "home-brewed" computer system from Hank Crook; you can trust me even though my brother is a scum who says that this computer contains a valid TPM inside. Now you don't trust him, along with every customer doing business with Hank, but he does have good pricing. Hank, on the other hand, knows his reputation isn't exactly the best, so he goes to a CA and gets X509 Certificates for all the TPMs resident within his computer systems. Now, when a customer of Hank asks, "Why should I trust you, your brother is scum?" Hank can prove that there is a valid TPM in the computer by looking up the certificate generated by a third party, in this case a CA trusted by the customer.

This is all well and good, but "How exactly does a X509 Certificate or any certificate get attached to the TPM?" Good question; the short answer is the TPM Endorsement Key. Remember, the EK is unique with regard to an individual TPM, and the EK public key is an identifier that can be used to ascertain its origin. Therefore, when the CA generates an X509 Certificate, the

EK public key is presented as proof of the TPM's identifier. When the certificate is generated and someone wants to know which TPM this certificate certifies, they read the public EK, which identifies the individual TPM and the certificate that belongs to that device. The point in all of this X509 discussion is that the TPM EK provides a root-of-trust regarding the ability to uniquely identify the TPM and to ascertain its origin. In some cases, the TPM vendor can produce the X509 Certificate, but the more likely scenario is that the TPM vendor gives information regarding the TPM that allows the OEM to produce an X509 Certificate. Figure 4.1 shows a typical scenario regarding the creation of an X509 Certificate by an OEM.

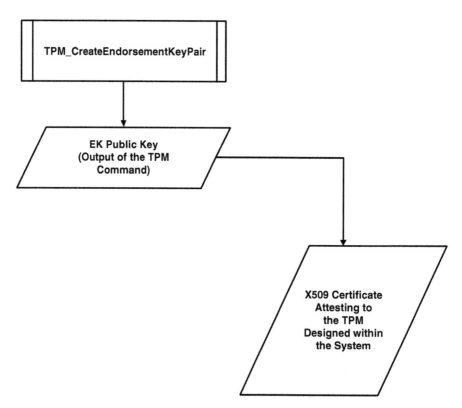

Figure 4.1 The creation of an X509 certificate by an OEM

In the environment of an embedded system, the idea of an X509 Certificate is usually moot. The more likely case is that the designer, or designers, of the system purchases the TPM from a trustworthy distributor who ensures that it is indeed an Atmel TPM. The real security issue for this type of TPM deployment resides within the EK cryptographic functionality and the root-of-trust it provides with regard to the host system's security.

4.4   Security and the EK

The EK is the root key in the TPM Key Management scheme; the next chapter explains the TPM Key Management concerns. This implies that the EK is the first RSA key pair created within the TPM and is used to support the creation of all remaining RSA key pairs that may exist, at some point in time, within the TPM. Therefore, any RSA key created within the TPM has a root-of-trust that is traceable to the EK. In other words, the trust is extended from the EK to all RSA key pairs that might be created in the course of TPM operation or the root-of-trust can be ascertained by tracing any RSA key within the TPM back to the EK.

Expanding this thought, the EK is attached to one and only one host system platform. The host system can have only one TPM per TCG specification, and the root-of-trust provided by the TPM EK extends to external as well as internal systems of the TPM. In other words, the host system knows the unique identity concerning the TPM residing on the system by virtue of the EK public key, and if this TPM were swapped for another, the host system would detect the EK public key mismatch. Not only does the host system know that the TPM was swapped, but it also knows that all of the remaining TPM keys are invalid, relative to this host system, because the root-of-trust has been modified. In other words, the host system is tied or bound to a specific EK; this is a cryptographic relationship between the host system and the root-of-trust, which extends trust to other entities within the TPM. Figure 4.2 shows the relationship between the host system and the TPM EK.

Another task associated with the EK involves supporting the establishment of a TPM owner. Remember, the only RSA key pair residing within the TPM prior to deployment, from the TPM manufacturer's point of view, is the EK; and in some cases, the OEM must generate the EK. The point is that the EK, at some place in time, is the only RSA key pair held within the TPM. The owner has two secrets associated with it: the Owner secret and the Storage Root Key (SRK) secret; more on these in later chapters. Secrets being secret, there has to be a mechanism to store these secret values within the TPM without devulging their content. The following describes how the EK comes into play.

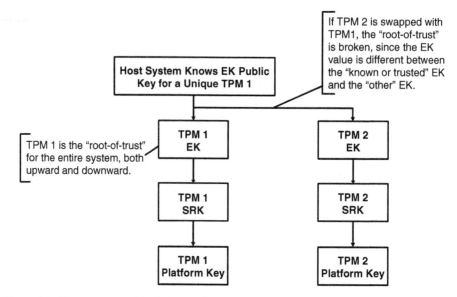

Figure 4.2 Host system to TPM EK root-of-trust

The EK is defined as a 2048-bit Storage Key or an RSA Encryption Key; and you are not allowed to perform an RSA signature using the EK. This means that the EK is available to an RSA to encrypt data – that is exactly what we are going to do. Therefore, to protect the secret information during input command-message transmission to the TPM via SMBus or LPC, we encrypt, separately, the Owner and SRK secrets using the EK public key. When the TPM needs to use these secrets, it simply decrypts the cipher text using the EK private key. Note that since the EK cannot leave the TPM, ever, only one TPM can successfully gain knowledge of these secrets. Figure 4.3 shows this concept in graphic form.

Granted, the explanation does not shed light on the specifics of the use of the decrypted secrets; this is not the intention of these early chapters. The point to take away from this discussion concerns the EK and its cryptographic ability to extend its root-of-trust to help create internal entities.

In summary, the EK provides a unique cryptographic identity with regard to a specific TPM. The EK is used to encrypt secrets used to establish a TPM owner. The EK is a 2048-bit nonmigratable RSA Encryption Key that never can be used for signature purposes. Host system platforms can bind the TPM by leveraging the EK public key and by extending the root-of-trust upward, as well as internally for entities in the TPM. The EK is created once

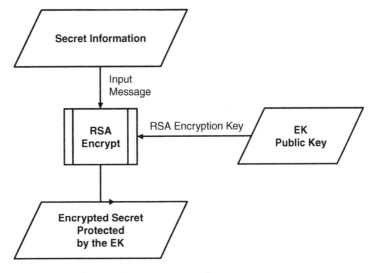

Figure 4.3 The transfer of secret information using the EK

and can never be removed from the TPM and must be generated within the TPM to maintain the private key anonymity. X509 Certificates can be generated to attest to the EK residing within the TPM. Embedded design engineers can leverage the EK cryptographic properties to establish a secure host system design by virtue of the root-of-trust inherent in the EK. The next chapter extends this idea by discussing various key hierarchy and management issues about the TPM.

Key Hierarchy and Key Management

With the discussion of the Endorsement Key (EK) behind us, we can now focus on the Trusted Platform Module (TPM) key hierarchy and the resulting key management schemes. One thing to keep in mind is the constant or fixed-key hierarchy that exists within every TPM, regardless of manufacturer. Other keys loaded into the TPM are of the managed type, meaning that the design engineer will have a particular key hierarchy in mind when designing a host system leveraging the TPM. It is the responsibility of the design engineer to define a key hierarchy scheme that satisfies the requirements of the system for the backup or archiving procedure, as well as the type and the ownership of keying material. Any one of these issues, when done without some preparation work concerning system operability, can cause major problems during system development or on system deployment. The goal of this chapter is to give the reader some sound advice regarding the methods and reasons involved in key hierarchy design along with corresponding key management functionality.

5.1   TPM-Specific Key Hierarchy

The Trusted Computing Group (TCG) specifies a minimum key hierarchy in regard to the EK, the "root-of-trust", and to the Storage Root Key (SRK) – the TPM Owner's cryptographic key. This foundational key hierarchy cannot be altered and every TPM-based design must follow the relationship specified concerning the EK, SRK, and keying material loaded within the TPM. Another way of saying this would be: the root-of-trust established by the EK must be extended to the TPM Owner and sequentially extended through the owner to all keying material held in the TPM. Figure 5.1 shows this concept as it relates the EK, SRK, and some abstract keys that might be loaded into the TPM.

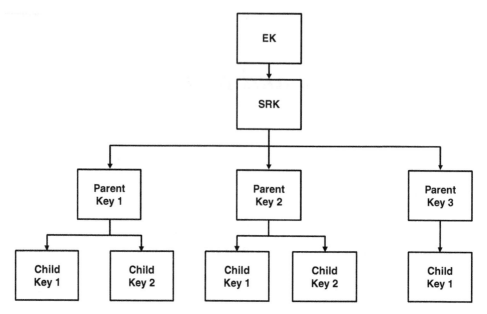

Figure 5.1 Example key hierarchy

One can see that the EK is cryptographically related to the SRK by virtue of the SRK secret being encrypted by the EK public key. Since the cipher text can only be decrypted by an EK private key whose existence is confined to a specific TPM, the root-of-trust is extended to the SRK. The SRK is the parent key to all keys below it and the SRK public key, like the EK, protects private data during the load procedure of each key that is a child of the SRK. Hence, the root-of-trust is extended from the EK to the SRK and through to the child keys directly below the SRK. In addition, each child key, relative to the SRK, has its own pair of child keys, which are loaded into the TPM under the cryptographic protection of its parent key. Thus, the root-of-trust is also extended to the child keys whose parent is itself a child key of the SRK. This example clearly shows that root-of-trust is extended from the EK through the entire key hierarchy to each of the lowest keys within it. The next section discusses which types and sizes of keys the TPM can support.

5.2   Types of Keys Found within the TPM

There are six types of keys that are allowed to be loaded into the TPM as defined by the type TCG_KEY_USAGE value. Figure 5.2 shows the possible values that can be equated to this type of identifier.

TPM_KEY_SIGNING	0x00	0x10
TPM_KEY_STORAGE	0x00	0x11
TPM_KEY_IDENTITY	0x00	0x12
TPM_KEY_AUTHCHANGE	0x00	0x13
TPM_KEY_BIND	0x00	0x14
TPM_KEY_LEGACY	0x00	oc15

Figure 5.2 TCG_Key_Usage types

Before we get into the specific types of key that can exist within the TPM, an explanation is in order concerning the RSA. An RSA key is capable of encrypting, decrypting, signing, and verifying signatures with the last two cryptographic functions being a form of encryption, private key, and decryption, public key. The TPM wraps the generic RSA cryptographic capabilities and forces the individual keys to be of a certain type defined by the key usage. Therefore, the TPM allows a TCG_KEY_SIGNING key to exist that is used specifically for signing various forms of data. The TPM also allows a **TCG_KEY_STORAGE** key; examples of this type of key are the EK and SRK, which are used to encrypt other keys within the key hierarchy.

Next is the **TCG_KEY_IDENTITY** key, for which use is limited to the TPM_Identity command, which is outside the context of a TPM basics book. The **TCG_KEY_AUTHCHANGE** key is another specific-purpose key related to the TPM_ChangeAuthAsym command, which is also outside of the TPM basics context. The final key of interest is the **TCG_KEY_BIND** key, which allows data to be bound or unbound from the TPM, logically, by a cryptographic encryption/decryption operation. Hey, wait a minute, there is one more type that wasn't mentioned – that's right, the **TPM_KEY_LEGACY** type. The Legacy key is a deprecated type that allows the key to sign and bind – encrypt. This is really an RSA key with no functional wrapper and is not recommended for use; the system designer knows, I would hope, which RSA operations need to be performed by the different keys held within the TPM. **TPM_KEY_LEGACY** was defined in the specification in response to companies that used this type of key within legacy cryptographic designs.

All of the keys mentioned are defined, with regard to their use, in later chapters of this book. The point here is to introduce the reader to the various key types available for use within the TPM during the further discussion concerning key management and hierarchy.

5.3   Typical PC-Based Key Hierarchy

First, let's look at the normal PC-based TPM key hierarchy to get acquainted with an easy-to-understand key management scheme. Figure 5.3 describes a simple PC-based key tree used by an information systems (IS) department setting up a single PC to service three individual employees.

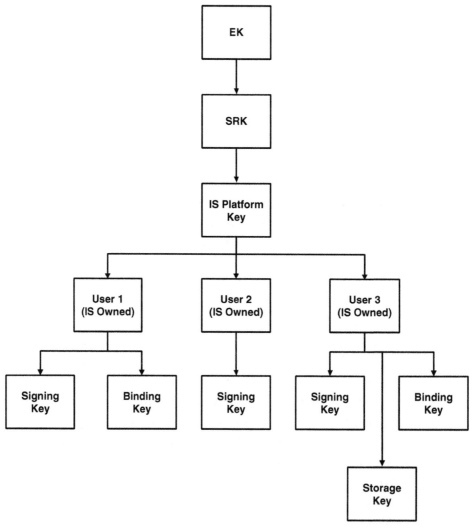

Figure 5.3 A simple PC-based key hierarchy

Notice our two old friends, the EK and SRK. Below the SRK, is an IS-owned key, which is named the "IS Platform Key"; from a key management point of view it is only accessible by the IS department. The next keys in the key hierarchy concern three user keys, which separate three individual user profiles and give each one access to the TPM. Finally, each user profile has a subset of keys available that give each specific TPM cryptographic capabilities.

First up is User 1. This person has a signing key and a binding key available to allow the user to leverage the TPM concerning signatures and encrypting data. That is all this individual user can do. For example, User 1 cannot load any new keys into the TPM since he or she does not own a storage key; remember, the storage key allows addition keys to be added to the key hierarchy. Some of you may be thinking that the User 1 key is itself a storage key and you would be right, but look at which entity owns the key. That's right, the evil IS department owns the key; sorry User 1, you lose.

Now User 2 must be really disliked by the IS department personnel – this profile only allows its user to perform a signature operation using the TPM. Again, User 2 cannot do anything about this since the IS department owns User 2's storage key and it is not about to give this loser any more freedom concerning the TPM.

Before we get to User 3, let's discuss how the IS department "owns" the user keys; by the way, this also includes User 3's storage key. When a key is loaded into the TPM, a Usage Secret is associated with that key and needs to be leveraged to obtain the cryptographic services defined by that key. IS simply keeps the User Secret hidden from our three PC users, which prevents them from leveraging the use of particular storage keys.

On to User 3 – this man is well liked within the IS department. User 3 has the same capability as User 1, with one exception; this lucky sap has a storage key. This means User 3 can load whatever key type he desires, using the storage key as a parent. In addition, there is no limit, other than TPM resources, to the depth of the user's key tree. For example, User 3 can load another storage key using the top storage key as a parent and can expand the key hierarchy. Sooner or later this user will want to perform a meaningful function using the TPM and User 3 will eventually realize that he should load a key to do something, like an RSA signature. The point is that User 3 has a much better deal regarding personal key management; for example, the person could choose to add a storage key below the second-tier storage key and expand the key hierarchy. Figure 5.4 shows this possibility with regard to User 3's key management choices.

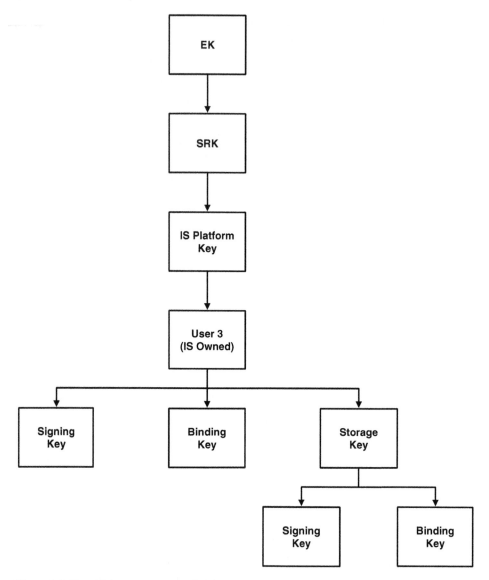

Figure 5.4 User 3's key management decision

5.4 Key Flags and Their Meaning

In addition to the key usage values there are various ways in which TPM key-ing material can be managed inside the TPM. One of these ways is with an additional parameter – the **TCG_KEY_FLAGS** – defined within the TPM; this is a masked value. The value may define one or all of the properties listed in the table shown in Figure 5.5.

No modification	0x00	0x00	0x00	0x00
Redirection	0x00	0x00	0x00	0x01
Migratable	0x00	0x00	0x00	0x02
volatileKey	0x00	0x00	0x00	0x04

Figure 5.5 TCG_KEY_FLAGS

The two key flags leveraged the most are the **Migratable** and the **volatileKey** flag bit masks. The first of these two mask values' functionality seems quite obvious, to allow the TPM key to be migrated from the TPM. This does mean that the key will come out of the TPM in the clear, only that the key can be removed from the TPM, securely, by cryptographic means. One purpose for this functionality concerns the archiving or backup of the keying material held within the TPM. Key material archiving is discussed later in this chapter.

The other mask value is the **volatileKey** bit; this value defines just that – the key is volatile concerning TPM power cycles. Now someone might say "What good is that, deleting a key on every TPM power cycle?" There is a rhyme to this reason and this involves the TPM resources with regard to the number of keys that can be held within the TPM at any given time. Most embedded designs will not come across this problem, but one never knows. In Figure 5.3, we saw three PC users who had access to keying material stored inside of the TPM and that User 3 has very liberal boundaries concerning the number of keys he can store within the TPM. What happens if User 3 tried to store a very large number of keys within the TPM for which there were insufficient resources to accommodate this type of action? The answer is that the TPM would eventually respond with an error response TCG_NOSPACE, indicating that the internal key heap was full.

One way to avoid this situation is to make the additional keys volatile, which would flush the keying material from the TPM on every power cycle. This would mean that the TCG Software Stack (TSS) – we are talking PC language right now – would realize that these keys were of the volatile type and reload the keys after power is reestablished. Now, someone could say "But the number of keys trying to be loaded is still too much for the TPM." This is true and the TSS would make the decision as to when any one key would have to reside within the TPM and leave some of the other keys outside its internal boundaries until they are needed. This would be accomplished by executing the TPM_EvictKey command, which would remove the key from the TPM's, internal storage. Also, by making the keys volatile, the TPM internal key heap is cleared of this extra keying material; therefore User 3 can be fairly inept when it comes to key management, putting the burden on the TSS, which chooses to reload the keys during the next system cycle. The bottom line is that keys can be made volatile and their scope can be bounded by an individual power cycle.

The final mask value, regarding key flags, concerns the **Redirection** flag. This flag will be discussed at the appropriate time when its use is defined within the context of actual command usage. The flag has an obscure use and its description will make more sense within the boundaries of that use.

5.5   Key Cryptographic Algorithm Definition

Another aspect of the TPM internal keys involves the cryptographic algorithm that these keys are associated with. Consider the TCG_ALGORITHM_ID value, which contains the possible algorithms supported by the TPM, as defined by the TCG. Figure 5.6 shows all of the possible values this type of value can have.

TCG_ALG_RSA	0x00	0x00	0x00	0x01
TCG_ALG_DES	0x00	0x00	0x00	0x02
TCG_ALG_3DES	0x00	0x00	0x00	0x03
TCG_ALG_SHA	0x00	0x00	0x00	0x04
TCG_ALG_HMAC	0x00	0x00	0x00	0x05
TCG_ALG_AES	0x00	0x00	0x00	0x06

Figure 5.6 Possible cryptographic algorithm IDs

The list is very descriptive and, for this discussion, the only applicable value concerning keying material is the TCG_ALG_RSA, which defines the RSA cryptosystem. All keys loaded within the Atmel TPM are of type RSA. The Atmel TPM does support the secure hash algorithm (SHA-1) and Hash-based Message Authentication Code (HMAC) functionality, but this does not affect the loading and use of any cryptographic keying material that can exist within this TPM. The moral of this story is: if you are loading keying material into the Atmel TPM, it will be of the type RSA.

One other parameter associated with the algorithm ID concerns the scheme, or how the algorithm will perform its cryptographic tasks. With regard to the RSA cryptosystem, the two operations of interest about the scheme are its encryption and signature functions. Addressing encryption first, the possible choices are listed in Figure 5.7 as to the RSA scheme used when encrypting or decrypting data or cryptographic ciphers, respectively.

TCG_ES_NONE	0x00	0x01
TCG_ES_RSAESPKCSv15	0x00	0x02
TCG_ES_RSAESOAEP_SHA1_MGF1	0x00	0x03

Figure 5.7 Possible encryption schemes

The three choices are none, encrypt/decrypt using RSA Public Key Cryptography Standards (PKCS #1 v1.5), and encrypt/decrypt using RSA PKCS #1 v2.0. This may seem like a confusing choice considering that you might not be PKCS savvy, but the choices can be made very easy. The first one is quite obvious: perform no encryption/decryption with this RSA key, which would be selected if the key were to be used for signing. The next choice involves the TCG_ES_RSAESPKCSv15 scheme value and is considered an older style of encryption scheme. This choice encodes the data using PKCS #1 v1.5 and then encrypts the encoded data with the RSA key.

The other choice for the RSA scheme is the TCG_ES_RSAESOAEP_SHA1_ MGF1 value. The encoding is defined as Optimal Asymmetric Encryption Padding (OAEP) and is specified within PKCS #1 v2.0. The point here is that this is not just an encryption process; the data must be altered prior to encrypting the data. This is done so that multiple encryptions of the same data using the same key results in different ciphers. Without the encoding, encrypting the same data multiple times would produce the same cipher text.

This book is not intended to explain the art of crypto analysis; at the end of this chapter, I list a couple good resources that explain the various attacks that can be posed on ciphers, which do not involve encoding. The moral of the story is to choose the TCG_ES_RSAESOAEP_SHA1_MGF1 value if in doubt about which scheme to use; this is a more current method regarding this function.

The last scheme that has to be dealt with concerning the RSA is the signature scheme. There are three choices for this scheme and the values are shown in Figure 5.8. Once again, there is a choice regarding no signature capability when using the RSA key that this modifier is attached to.

The TCG_SS_NONE value is used if the key is to be of the signature flavor, which prevents the key from encrypting/decrypting data. Now someone

TCG_SS_NONE	0x00	0x01
TCG_SS_RSASSAPKCSv15_SHA1	0x00	0x02
TCG_SS_RSASSAPKCSv15_DER	0x00	0x03

Figure 5.8 Possible signature schemes

might say, "Well, I will just set the RSA key schemes to both encrypt and sign and be done with it." Wrong, the TPM will not allow this, and if you try to load a key with both schemes set to sign and encrypt, the TPM will refuse to load one. One exception concerns the Legacy Key type, which is a deprecated type and allows this configuration scheme; however, I don't recommend it. Not using the Legacy Key puts a burden on the key management design, but having a key that can sign and encrypt data can cause security problems, which is discussed in Chapter 17 about the TPM_Sign command.

The next RSA signature scheme is the value TCG_SS_RSASSAPKCS1v15_SHA1 and is fairly easy to understand. What this scheme requires is a SHA-1 digest to sign. For example, if you want to sign a message using this scheme, you would first SHA-1 hash the message then sign the resulting 20-byte digest. One benefit in using this signature scheme is that you are not limited by the message size; the message has to be reduced to 20-bytes via the SHA-1 function prior to signing. Figure 5.9 depicts this procedure.

Finally, the other choice of signature schemes is the TCG_SS_RSASSAPKCS1v15_DER value. This choice allows for the signing of a raw message – one that has not been hashed. There are some limitations with regard to this scheme's usage, which involve the size of the message that can be signed. To understand this, one must know the basics regarding signature size versus key size. If the RSA key is 2048, 1024, or 512 bits, the resulting signature will be 256, 128, or 64 bytes. Now someone might say, "Good, I can sign a message that is 256, 128, or 64 bytes in length." Wrong, there is a little problem called encoding, specifically Distinguished Encoding Rules (DER) encoding. What this means is that the DER encoding takes up a certain number of bytes from the available byte pool associated with the RSA key size. The resulting message length that can be signed when DER encoding is applied is the maximum signature size minus the DER pad or 11 bytes. This makes the total message size that can be signed with a 2048, 1024, or 512 bit key 245, 117, or 54 bytes in

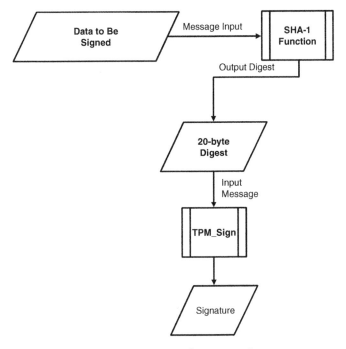

Figure 5.9 Signing a message using the SHA-1 scheme

length. The TPM will handle the DER encoding; all that has to be supplied is the message length. More on this subject is in later chapters.

5.6   Putting It All Together

OK, let's put this all together with regard to key management. Figure 5.10 shows a possible key management structure that may exist within the TPM leveraging all of the keying material components discussed so far.

Notice that the EK and SRK are present along with a platform key; this key is migratable and will allow the keys defined under it to be migrated off the TPM for archival purposes. Instead of the PC version, this figure uses application-based keys, which define TPM resources that applications can make use of thereby limiting each application to specific cryptographic capabilities. For example, within the embedded system there might be an application that is tasked with the encryption and decryption of data. There would be no need for this application to gain access to the signature capability of the TPM; hence, the no signature keying material would be associated with this application. Note that if a Legacy Key was associated with this

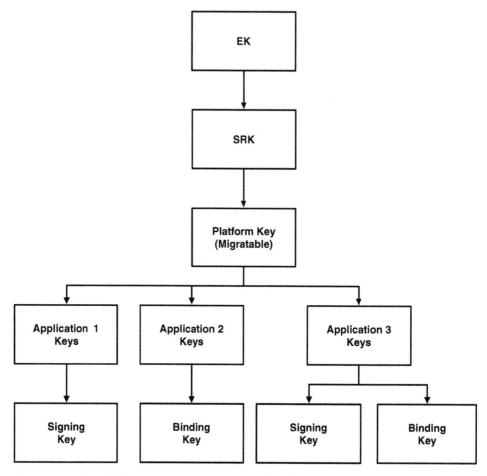

Figure 5.10   Possible key management structure within the TPM

particular application, the application could inadvertently sign data attesting to absolutely nothing; this is an encryption key. All other applications are given various accesses to cryptographic capabilities made available by the TPM depending on their needs.

5.7   Key Migration and Archiving

Imagine that you designed an embedded system using great key management architecture and deployed the system with a pat on your back. Six months later, customers are calling stating that the system failed and they can't get it up and running again. With a little investigation you find out that the system has "a little bug" – it flushes all the keying material from the

TPM, which is why all the systems are failing. Oops, you forgot to design key archive support into the key management solution. Don't let this happen to you. Encrypted data protects content from malicious individuals; you just joined that group if the key used to encrypt data is lost.

With that said, let's look at a possible key management solution that can help retrieve internally stored TPM keying material. Refer to Figure 5.10, and notice that the key under the SRK – the Platform Key. Now look closely at its properties; see anything in particular? This key is migratable. Further investigation will reveal that most of the keys under the Platform Key are migratable. The situation that we have here involves the parent key being migratable, allowing it to be stored outside the TPM. Since the Platform Key is migratable and it is the parent of all the keys below it, all the keys may be migrated off the TPM as well. Now the parent of the Platform Key is the SRK associated with an individual TPM. Once we have the Platform Key outside the TPM, we can migrate all the keys, the Platform Key along with all the child keys, to another TPM. By doing so, the Platform Key will have a new parent – the SRK associated with the other TPM that it was migrated to.

Basically, we rewrap the Platform Key from one SRK parent to another SRK parent; remember, the SRK protects the private portion of its child key. Technically, three rewraps are being performed: the rewrap of the Platform Key and its child keys to the protection of an external archive key, then another rewrap from the archive key to the new SRK. Figure 5.11 shows this relationship. The specific command suite that supports the migration of keying material held within the TPM is described in Chapter 21 on TPM key migration.

Now, back to our problem concerning the system bug that has flushed the keying material from customers' systems. If a key archival scheme was designed into the key management of the systems in question, the design engineer or customer service rep could rewrap the backup system keys with the each system's TPM SRK and reload the keying material. The moral of the story is: Don't paint yourself into a corner thinking about how secure you can make a system and, in doing so, forget that systems can and will fail. Design in key backup or archival capability so that when the system does fail with regard to keying material or even the TPM, God forbid, you are covered.

In summary, we have seen that the TPM has a fixed root-of-trust in regard to the EK and SRK. In addition, the TPM supports the creation of various RSA key types, which allow the system designer or security architect to design a robust key hierarchy realization. TPM key management is very

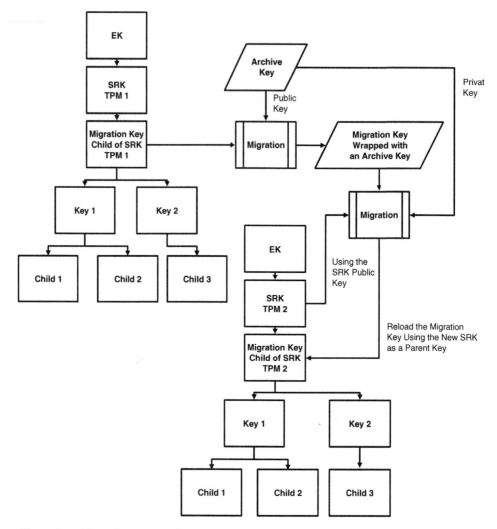

Figure 5.11 Key archiving procedure

important not only with regard to security, but also with regard to system keying material backup.

References

Bruce Schneier (1996). *Applied Cryptography: Protocols, Algorithms, and Source Code in C, 2nd Edition.* New York: John Wiley & Sons.

Douglas R. Stinson (2002). *Cryptography: Theory and Practice, 2nd Edition.* Boca Raton, FL: Chapman & Hall/CRC.

Platform Configuration Registers

One of the issues concerning the validation of any host system has to do with detecting changes made to that system after deployment. Just ask the design team of Microsoft's Xbox how they felt after the system was hacked in a way that allowed a Linux operating system (Linux OS) to exist on their "really cheap" platform. The purpose of the Xbox was to make the hardware cheap so that individuals could spend gobs of money on games to play using the device. In a perfect world, no one would be able to alter the hardware configuration and the system would operate as planned to allow access to the entire suite of cool games written for that platform. But, the designers "poached" the system security design and left the preverbal barn door wide open for hackers to create a really cheap version of a personal computer (PC) running a flavor of Linux. Not to mention that the Linux OS is free and is in direct competition with Microsoft's commercial versions of PC OS. But I can tell that the readers of this particular book are much smarter than that and are not going to let themselves get "powned" by some hacker with way too much time on their hands. Hence, the discussion regarding the Trusted Platform Module (TPM) and the Platform Configuration Registers (PCRs) contained within this device.

6.1 What in the World Is a Platform Configuration Register?

In short, the PCR is a TPM internal register that contains a 20-byte digest that represents a host system's software and hardware configuration metric. This metric can be set before system deployment and monitored throughout the system's lifetime with regard to any change in the hardware and software configuration. If the system detects a change in its configuration – a change in the corresponding PCR – the system can take action. This may take the form of shutting system operability down or limiting system functionality, right up to notifying the manufacturer that the system has been hacked. The point is that the system has the capability to protect itself from malicious attacks on it.

But what exactly is a PCR from a physical and logical standpoint, and how does one set the values within the internal TPM registers. First, let's discuss the number of PCRs available in a given TPM along with host system properties that each group of PCRs is associated with. As stated, a PCR register can protect either a host system's hardware or software configuration and there are 16 PCRs available for use within the TPM, 8 for hardware and 8 for software. Figure 6.1 describes the two groups of PCRs related to both the hardware and software of the host system configuration.

Hardware PCRs	Software PCRs
PCR 0	PCR 8
PCR 1	PCR 9
PCR 2	PCR 10
PCR 3	PCR 11
PCR 4	PCR 12
PCR 5	PCR 13
PCR 6	PCR 14
PCR 7	PCR 15

Figure 6.1 Hardware and software PCR grouping

The 16 PCRs in each TPM monitor the configuration of 8 hardware and 8 software host system capabilities and detect changes in each of the entities. Let's take a simple example: the notion of PCRs attesting to host system configuration and modification thereof. In Figure 6.2, the host system configuration is protected by two PCRs, one concerns the system

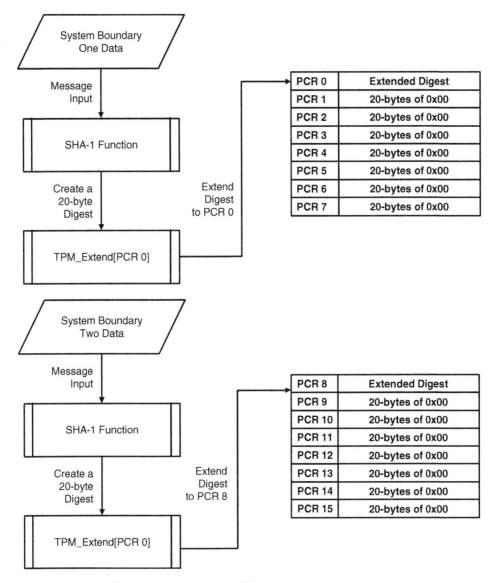

Figure 6.2 Simple host system leveraging two PCRs

boot-up Flash and the other concerns a single application running on this system.

Notice that the boot-up Flash region is considered to be part of the host system hardware configuration and the application is considered to be part of the host system software configuration. The two resulting regions that we are concerned with are the physical boundaries containing the entity we

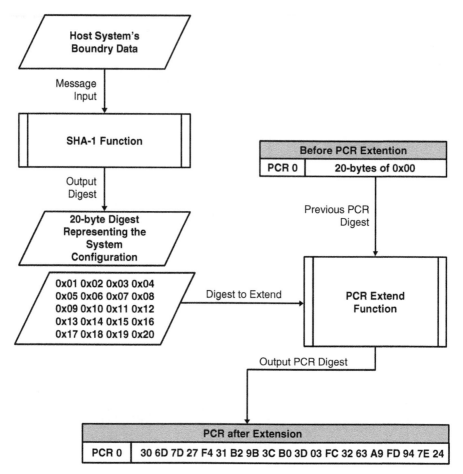

Figure 6.3 PCR extension algorithm

wish the monitor regarding alterations. The PCRs are extended by hashing the area defined by the physical boundaries associated with each of the entities to be monitored. For example, the bootstrap code exists in Flash memory between address locations 0x0000 and 0x1000. PCR 1 is assigned to this boundary and the data existing between 0x0000 and 0x1000 is hashed and the resulting 20 bytes are extended to PCR 0. In addition, the application code's physical address space is bounded by address locations 0x8000 and 0xF000; this area is hashed and the resulting 20 bytes are extended to PCR 8. All of the resulting PCR digests are then stored within the host system, preferably in a hardware register.

Now, when the system is in the field and every time the system boots, one of the first tasks is to compare the "on-the-fly" hash digest, extended to the two PCRs, with regard to each region, to the digest stored in the host system of PCR 0 and PCR 8. If the digests match the corresponding PCR values, the system boots as normal; but if one or both of the digests disagree(s) with either PCR 0 or PCR 8, the system takes preventive action. Care must be taken when designing an embedded system that makes use of this type of host system configuration management; what, you think because you have a TPM on board you don't have to worry about security? Wrong!

For example, if the digest check is performed in a soft-based manner, code running out of Flash for instance any hacker worth their salt can replace the on board Flash and successfully modify the system by altering the digest verification routine. The point is to make the reader aware that host system security does not end with the TPM and security must be done correctly during host system design; if you don't believe this, ask Microsoft.

6.2 How PCR Values Are Initialized

Well all of this PCR stuff sounds good, but how does a PCR get extended or initialized so to speak? Well, to answer this question we need to look at Figure 6.2 again and notice that all PCR values are initialized to zero prior to the host system's digest calculation and resulting PCR extensions. There are a few more TPM requirements that will be discussed in the appropriate chapter about the TPM PCR initialization rules. The algorithm used for the PCR extension is defined in Figure 6.3.

The host system creates a 20-byte digest by SHA-1 hashing the area data to be monitored with regard to configuration. Once this digest is available, it, along with the PCR index, is passed to the TPM in the form of the TPM_Extend command; when this command successfully executes, the PCR value has been modified accordingly. Since the PCR digest value is comprised of all zeros, on system boot-up the PCR will contain the initial digest representing the system configuration being monitored. By referencing the PCR Extension algorithm this result can be proven; I leave that task as an exercise for the reader. Using this procedure, each of the 16 available PCRs can be extended and initialize each PCR with the appropriate digest.

OK, we have extended our PCRs and the host system performs configuration checks during each and every power-on state. What happens if the system needs to be modified in lieu of a system modification driven by the manufacturer? This is a good question since this type of procedure happens quite often in the manufacturing world that drives us all nuts. You go to your favorite geek shop and purchase that latest new thing, only to come home, fire it up, and find out that the system firmware needs updating because the device was shipped with buggy code. No, this never happens in the real world. Well, if your deployed system checks its configuration and governs system operability based on this metric, how does one update to a new valid configuration in the field? The answer is the same procedure that was used when the system was manufactured. In the preceding example, let's say there was an update to both bootstrap and application code and the manufacturer wants to not only modify the deployed systems but also to modify the system configuration check.

First, the upgrade utility has to modify each PCR digest in reference to host system boundaries' new data, the bootstrap routine and the application code. To maintain the revision history about previous hardware and software states that have existed within the system prior to each and every update, the system must take into account the current system metric before updating to a new system configuration. This means that the PCRs will have to reflect not only the configuration resulting from the update, but also the previous configuration prior to said update. What this requires is the extension of the PCR(s) digests from the state that the system is currently in to the configuration that the system is going to, while keeping a history of previous system configurations. So instead of extending the PCRs relative to a zero-based value, when the system was manufactured, the system PCRs need to be extended relative to their current values. Figure 6.4 describes this concept.

The first task the update utility must do is to calculate the SHA-1 hash regarding the data that will be used to update each system component. This will result in a 20-byte digest, which represents the updated system configuration. Next, the update utility must concatenate the last system digest with the new system digest and SHA-1 the resulting 40-byte message. The result is a new system metric for that particular system entity; in this case, the bootstrap code that reflects not only the new system configuration but also the system configuration that existed previously. The system must still keep a record of the "old" digest, with regard to the bootstrap code's previous configuration, along with the current digest. These two digests allow the system, on power-up, to successfully test the new system configuration relative to its previous configuration.

Before the new configuration can be tested at power-up, the TPM must also reflect the changes applied by the system update. This is done simply by

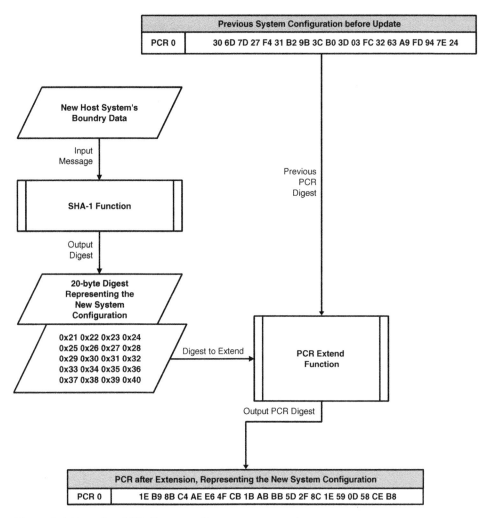

Figure 6.4 Extending PCR digests to reflect a system upgrade

extending the PCR digest, in the case of the bootstrap code PCR 0, with the new system digest described earlier. Now PCR 0 reflects the new system configuration of the update to the host system. Therefore on power-up, the system will hash the new bootstrap code and concatenate the previous digest with the digest of the new bootstrap code. Then, SHA-1 hashes this value and resulting digest must match the digest held within PCR 0 in order for the system to boot properly; otherwise, the system will fall into an error routine. Figure 6.5 depicts this system boot functionality.

Expanding on our host configuration monitor routine, the system must keep track of two digests: the current system digest and the previous system

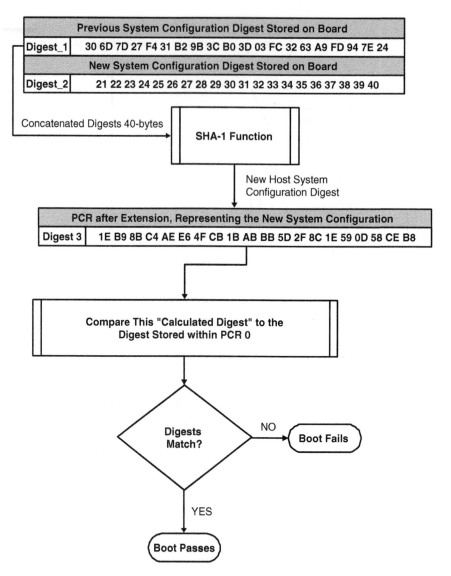

Figure 6.5 Testing the new host system's bootstrap configuration

digest for each boundary that is to be tested. This can mean up to 32 digest values if all 16 PCRs are to be leveraged. To be clear, the previous system digest would be initialized to a zero value at the time of manufacturing deployment, since the current system digest reflects the initial system configuration. The previous system digest must be present to hold the last system configuration in order to maintain the accumulative system update history. Therefore, every time the system is updated, the last system configuration digest would be modified by SHA-1 hashing the current system data, concatenating the last system configuration digest with this new digest, and SHA-1 hashing this 40-byte message. The last system configuration value will be assigned to the digest just described in order to reflect the system configuration prior to the new system update. This procedure will happen every time the system is updated, extending the last system configuration digest to reflect the current system configuration prior to updating.

6.3   How PCRs Govern TPM Command Execution

The preceding scenario is an example to help readers understand PCR digest usage within the context of the host system. The real interesting subject regarding PCR digests concerns how the TPM will leverage these configuration metrics. First, instead of the host system enforcing configuration metrics, defined by PCR digests, we will leverage the TPM to enforce command execution based on these metrics. This is done by tying or binding specific TPM entitles, specifically keying material internal to the TPM, to PCRs. The results allow the TPM to deny specific command execution, referring to PCR-bound keying material, based on predicted host system configuration or lack thereof.

The keying material is bound to PCR digest(s) during TPM_Create Wrap Key and TPM_LoadKey command execution; the PCR digest value(s) are resident at the time of keying material creation and subsequent loading. This states that the configuration of the host system is taken into account when the keying material is created and loaded. Therefore, if the host system's configuration has been modified, the created and loaded keys will be useless with regard to the command execution that leverages the entities. This idea is defined by the concept of having keying material bound to a selection of PCR digests, which define host system configuration at the time when the key was created or loaded within the TPM. Figure 6.6 describes this concept by showing a key that has been created during a specific host system configuration and a load operation's execution, using this same key, in a modified host system configuration.

One other aspect concerning PCR binding is with regard to the TPM_TakeOwnership command and the requirement of this command to generate a Storage Root Key (SRK). The SRK can also be bound to PCRs, as regards command input message block parameters, and for the use of this key, can be bound to a specific host system configuration. Much more detail

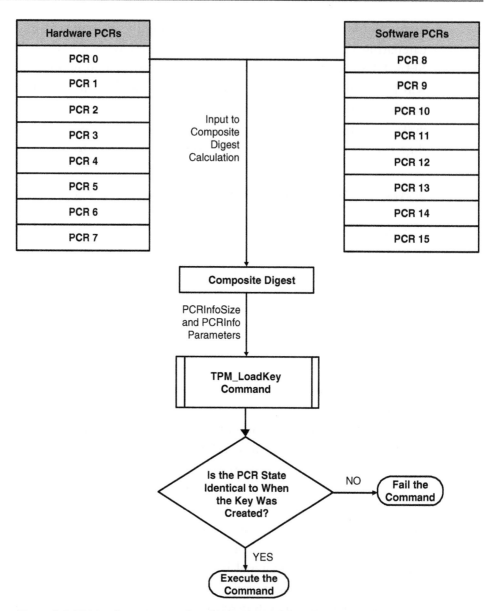

Figure 6.6 TPM enforcement regarding keying material bound to PCRs

regarding the specific input message block creation is included within the chapters that deal with taking TPM ownership and the key management command suite.

The host system is still responsible for supplying or extending the PCR digests, which will then be used by the TPM to enforce its command execution based on PCR digest content. Therefore, it is up to the host system's design team to come up with a solution regarding digest(s) that represents system configuration state. The digest generation scheme mentioned before is certainly appropriate concerning this task, but if the only metric of interest is current system configuration, the last system configuration digest(s) can be excluded from the equation. Then the system would only have to extend the current system metrics in an effort to produce a PCR digest that represents the current configuration. Since no previous system configuration would be taken into account using this model, there would be no need to support the additional 16 digests that represent a history of previous host system configurations.

6.4   Other PCR Tidbits

Extending the PCR values is great, but what if I want to simply read a PCR value? This can be done as well; any PCR can be read by sending a TPM_PcrRead command to the TPM indexing the PCR digest that is to be read. Just for edification, the security regarding host system configuration is directly related to the security provided by the current and last system configuration digests held within said system. The PCR is just a mechanism in which to enforce the use of the TPM facilities in reference to the current system configuration and not to secure the system's configuration digests. The moral of this story: be very careful regarding PCR digest creation and storage. Do not let the digests fall into the wrong hands or bad things will happen (see the earlier discussion about Microsoft).

Now we have talked about binding a PCR digest to specific TPM keying material entities, for example the SRK, but what if I would like to bind numerous PCR digests to a particular entity? Well here is a good time to discuss composite digests; this is a 20-byte digest based on the inclusion of individual PCR(s) selected by host system configuration interests. In other words, multiple PCR digests can be selected and reduced to a single digest that represents the host system configuration attested to by each individual PCR digest selected. Keying material that is being created or loaded into a TPM can be bound to multiple PCR digest values, and if host system configuration changes, measured by these individual PCR(s) and reflected in one digest – the composite digest – the specific command execution will be denied. This discussion concerns the overall concept of composite digests related to PCRs;

see the specific chapters concerning the Key Management Command Suite (15) and the PCR Command Suite (19) for more detailed descriptions.

One more tidbit of information regarding the PCR digests' contents and the clearing of these values during a TPM_Post initialization: this simply means a successful TPM boot-up sequence concerning startup and the execution of the internal TPM self-test suite. Depending on the type of TPM startup performed during host system boot-up, the PCR digests – the digests existing prior to power down – can be restored without recalculating the values (see Chapter 23 for more on the TPM Initialization Command Suite).

7

TPM Command Message Overview

In this chapter, the Trusted Platform Module (TPM) Command Message Blocks are defined, in a generic fashion, in order to understand the various forms of TPM command input and output messaging. The TPM has three levels of command authorization; no authorization, single authorization, and two authorizations. The level of authorization depends on the number of TPM entities referenced within the command message being sent to the TPM. All TPM command input and output messages have a common "header" and, depending on the command, will add information to the respective headers.

7.1 Non-authorized TPM Command Messages

The first command input and output generic message blocks examined here concern the non-authorized TPM command category. These types of messages do not reference any particular type of TPM entity – for example a TPM key – and need no authorization data attached. This doesn't mean that the command will be successfully executed by virtue of correct input message compilation; there is a little thing called TPM state, and a time and place regarding command execution depends on such state. Regardless of TPM state, the requirement of this discussion is to introduce the compilation regarding non-authorized TPM commands, so let's get cracking.

First off, we must define how the TPM determines the level of authorization that any given command aspires to. This is conveyed by the message parameter defined as the Authorization Tag and the possible values this type of parameter can hold are defined in Figure 7.1.

TCG_TAG Values	
TCG_TAG_RQU_COMMAND	0xC1
TCG_TAG_RQU_AUTH1_COMMAND	0xC2
TCG_TAG_RQU_AUTH2_COMMAND	0xC3
TCG_TAG_RSP_COMMAND	0xC4
TCG_TAG_RSP_AUTH1_COMMAND	0xC5
TCG_TAG_RSP_AUTH2_COMMAND	0xC6

Figure 7.1 Possible authorization tag values

Concerning non-authorized commands, the possible authorization tag values of the input and output command message blocks are **TPM_TAG_RQU_ COMMAND** and **TPM_TAG_RSP_COMMAND**, respectively. Given this information, let's look at a typical non-authorized TPM command input message block, as defined in Figure 7.2.

Authorization Tag	0x00	0xC1		
Parameter Size	0x00	0x00	0x00	0x0C
Ordinal	0x00	0x00	0x00	0x99
Startup Type	ST1	ST2		

Figure 7.2 Non-authorized TPM_StartUp input message block

The generic header information contains the authorization tag, the parameter size, and the command code. All TPM input message blocks, regardless of authorization level, have this header information in common. The only other parameter in the command-specific input message block concerns the startup type, which tells the TPM what state to startup in. This parameter is considered part of the input command's "payload" and is simply defined as information the command needs to successfully execute the operation requested. The command is transmitted to the TPM via the LPC Bus or SMBus and the TPM returns an output message block, indicating the result of the command execution.

Authorization Tag	0x00	0xC4		
Parameter Size	0x00	0x00	0x00	0x0A
Return Code	0x00	0x00	0x00	0x00

Figure 7.3 TPM_StartUp output message block

In response to the TPM_StartUp command input message block, the TPM returns a command output message block as defined in Figure 7.3.

The generic header information contains the authorization tag, parameter size, and the command return code. All TPM output message blocks, regardless of authorization level, have this header information in common. The return results concerning the TPM_StartUp command contain the simplest of output message block parameters, the generic TPM header itself, an authorization tag, parameter size, and return code. This is due to the fact that the TPM_StartUp has a very simple command response – the command was successful or the command had some type of error. If the non-authorized command returns any type of payload, the corresponding payload would be attached to the output message header and the Parameter Size would be adjusted accordingly.

7.2 Single Authorized TPM Command Messages

The next type of TPM command input and output message blocks concern the single authorized flavor. The two authorization tags that indicate this type of authorization level are defined as TPM_TAG_RQU_AUTH1_COMMAND and TPM_TAG_RSP_AUTH1_COMMAND, respectively, as shown in Figure 7.1. Given this information, let's look at a typical Single Authorized TPM command input message block as defined in Figure 7.4.

Notice that the command input message block has an authorization tag defined as TPM_TAG_RQU_AUTH1_COMMAND and, in addition, the command input message block has attached an authorization block to it. This book will separate the command-specific input and output message blocks, which contain a generic header and any possible command payload from any attached authorization block(s) when discussing command messaging as a whole. This makes the discussion concerning the authorization protocol,

Authorization Tag	2 bytes	0x00	0xC2		
Parameter Size	4 bytes	0x00	0x00	0x02	0x6a
Ordinal	4 bytes	0x00	0x00	0x00	0x20
Key Handle	4 bytes	0x40	0x00	0x00	0x00
TCG Key Structure	555 bytes	0x01	0x01	0x00	0x00
		0x00	0x10		
		0x00	0x00	0x00	0x02
		0x01			
		0x00	0x00	0x00	0x01
		0x00	0x01		
		0x00	0x03		
		0x00	0x00	0x00	0x0c
		0x00	0x00	0x80	0x00
		0x00	0x00	0x00	0x02
		0x00	0x00	0x80	0x00
		0x00	0x00	0x80	0x00
		0x00	0x00	0x01	0x00
		Encrypted Blob Reference Key Handle			
		0x00	0x00	0x01	0x00
		Encrypted Blob Reference Key Handle			
Auth Handle	4 bytes	0x00	0x00	0x00	0x00
nonceOdd	20 bytes	NO1	NO2	NO3	NO4	NO5
		NO6	NO7	NO8	NO9	NO10
		NO11	NO12	NO13	NO14	NO15
		NO16	NO17	NO18	NO19	NO20
Continue Auth Session	1 byte	0x00			
Authorization Digest	20 bytes	AD1	AD2	AD3	AD4	AD5
		AD6	AD7	AD8	AD9	AD10
		AD11	AD12	AD13	AD14	AD15
		AD16	AD17	AD18	AD19	AD20

Figure 7.4 TPM_LoadKey input message block

described in later chapters, easier to understand. The TPM_LoadKey command input message block has the usual generic header along with a huge command-specific payload. The payload is described as a generic block; the specifics will only confuse the issue concerning the topic of generic message block discussion. Additionally, the authorization block is tied to the command input message block by the authorization protocol and is reflected in the authorization digest – the last 20 bytes contained within the block itself.

For more information regarding authorization, please see the appropriate chapter(s) about this subject matter. Again, this command is transmitted to the TPM via the LPC Bus or SMBus and the TPM returns an output message block, indicating the result of the command execution.

In response to the TPM_LoadKey command input message block and attached authorization block, the TPM returns a command output message block with an attached authorization block as defined in Figure 7.5.

Authorization Tag	2 bytes	0x00	0xC5			
Parameter Size	4 bytes	0x00	0x00	0x00	0X38	
Return Code	4 bytes	0x00	0x00	0x00	0X00	
Key Handle	4 bytes	0x40	0x00	0x00	0x00	
Auth Handle	4 bytes	0x00	0x00	0x00	0x00	
nonceEven	20 bytes	NE1	NE2	NE3	NE4	NE5
		NE6	NE7	NE8	NE9	NE10
		NE11	NE12	NE13	NE14	NE15
		NE16	NE17	NE18	NE19	NE20
Continue Auth Session	1 byte	0x00				
Authorization Digest	20 bytes	AD1	AD2	AD3	AD4	AD5
		AD6	AD7	AD8	AD9	AD10
		AD11	AD12	AD13	AD14	AD15
		AD16	AD17	AD18	AD19	AD20

Figure 7.5 TPM_LoadKey output message block

The command output message block contains the **TPM_TAG_RSP_AUTH1_ COMMAND** authorization tag, which indicates the output message block has a single authorization block attached. This parameter is part of the generic header that is indicative of all output message blocks. In addition, the output message block includes a command-specific payload, which in the case of TPM_LoadKey command is comprised of a Key Handle. There is also an attached authorization block that attests to the validness of the output command message block with respect to the input message block associated with the command invocation. To be clear, every authorized command input has a corresponding authorized command output, with an exception. If the command execution fails, the TPM will respond with a non-authorized output message block indicating the error condition.

7.3   Dual Authorized TPM Command Messages

Finally, there are the dual authorized TPM input and output command message blocks. The two authorization tags indicate that the types of authorization levels are defined as TPM_TAG_RQU_AUTH2_COMMAND and TPM_TAG_RSP_AUTH2_COMMAND, respectively, as defined in Figure 7.1. Given this information, let's look at a typical dual authorized TPM command input message block as shown in Figure 7.6.

Authorization Tag	2 bytes	0x00	0xC3			
Parameter Size	4 bytes	0x00	0x00	0x01	0x84	
Ordinal	4 bytes	0x00	0x00	0x00	0x0C	
Key Handle	4 bytes	0x00	0x00	0x00	0x01	
Protocol ID	2 bytes	0x00	0x04			
EncAuth	20 bytes	EA1	EA2	EA3	EA4	EA5
		EA6	EA7	EA8	EA9	EA10
		EA11	EA12	EA13	EA14	EA15
		EA16	EA17	EA18	EA19	EA20
Entity Type	2 bytes	0x00	0x03			
Encrypted Data Size	4 bytes	0x00	0x00	0x01	0x00	
Encrypted Data	256 bytes	Encrypted Data				
Authorization Handle	4 bytes	0x00	0x00	0x00	0x00	
nonceOdd	20 bytes	NO1	NO2	NO3	NO4	NO5
		NO6	NO7	NO8	NO9	NO10
		NO11	NO12	NO13	NO14	NO15
		NO16	NO17	NO18	NO19	NO20
Continue Auth Session	1 byte	0x00				
Authorization Digest	20 bytes	AD1	AD2	AD3	AD4	AD5
		AD6	AD7	AD8	AD9	AD10
		AD11	AD12	AD13	AD14	AD15
		AD16	AD17	AD18	AD19	AD20
Authorization Handle	4 bytes	0x00	0x00	0x00	0x01	
nonceOdd	20 bytes	NO1	NO2	NO3	NO4	NO5
		NO6	NO7	NO8	NO9	NO10
		NO11	NO12	NO13	NO14	NO15
		NO16	NO17	NO18	NO19	NO20
Continue Auth Session	1 byte	0x00				
Authorization Digest	20 bytes	AD1	AD2	AD3	AD4	AD5
		AD6	AD7	AD8	AD9	AD10
		AD11	AD12	AD13	AD14	AD15
		AD16	AD17	AD18	AD19	AD20

Figure 7.6 TPM_ChangeAuth input message block

This command input message block has an authorization tag defined as TPM_TAG_RSP_AUTH2_COMMAND, which indicates that the message block has two authorization blocks attached to it. The TPM_ChangeAuth, again, has the standard header information, indicating the authorization level, parameter size, and command ordinal. One side note, just in case you are curious: the parameter size includes the command input message block size and the authorization block size. In addition, the authorization block size, referencing each of the input authorization blocks, is 45 bytes each. Therefore, this dual authorized command has 90 bytes for the two attached authorization blocks. The TPM_ChangeAuth command, like the TPM_LoadKey command, has a fairly sizable payload that is discussed in detail in the command-specific chapters. When this command is transmitted to the TPM via the LPC Bus or SMBus, the TPM must authorize both entities referenced within each of the input authorization blocks.

In response to the TPM_ChangeAuth command input message block, the TPM will respond with a dual authorized command output message block, as described in Figure 7.7.

Authorization Tag	2 bytes	0x00	0xC5			
Parameter Size	4 bytes	0x00	0x00	0x01	0x60	
Return Code	4 bytes	0x00	0x00	0x00	0x00	
Out Data Size	4 bytes	0x00	0x00	0x01	0x00	
Out Data	256 bytes	Modified Encrypted Entity				
nonceEven	20 bytes	NE1	NE2	NE3	NE4	NE5
		NE6	NE7	NE8	NE9	NE10
		NE11	NE12	NE13	NE14	NE15
		NE16	NE17	NE18	NE19	NE20
Continue Auth Session	1 byte	0x00				
Authorization Digest	20 bytes	AD1	AD2	AD3	AD4	AD5
		AD6	AD7	AD8	AD9	AD10
		AD11	AD12	AD13	AD14	AD15
		AD16	AD17	AD18	AD19	AD20
nonceEven	20 bytes	NE1	NE2	NE3	NE4	NE5
		NE6	NE7	NE8	NE9	NE10
		NE11	NE12	NE13	NE14	NE15
		NE16	NE17	NE18	NE19	NE20
Continue Auth Session	1 byte	0x00				
Authorization Digest	20 bytes	AD1	AD2	AD3	AD4	AD5
		AD6	AD7	AD8	AD9	AD10
		AD11	AD12	AD13	AD14	AD15
		AD16	AD17	AD18	AD19	AD20

Figure 7.7 TPM_ChangeAuth output message block

Again, the authorization tag, within the command output message header, contains the value **TPM_TAG_RSP_AUTH2_COMMAND** indicating that the command execution output message has two authorization blocks attached to it. Each of these authorization blocks must be validated separately to attest to the legitimacy of the message. As with the single authorized command, the output of a failing dual authorized command will be a non-authorized output message block indicating the execution failure mode regarding the command specifics.

In conclusion, the input and output message blocks can consist of three types of authorization levels: non-authorized, single authorized, and dual authorized. The authorization tag contained in the input and output command message block is directly related to the number of authorization block(s) attached. If an authorized command fails, the command output message will be a non-authorized message stating the nature of the command execution failure. This failure can be related to TPM state or lack thereof, specific command execution failure or authorization failure. The authorization levels are defined by the number of entities referenced by the TPM command about to be executed and knowledge of the entities Usage Secret must be known with regard to successful command authorization.

Rolling Nonces and Anti-replay Protection

Anti-replay? Rolling nonces? The Trusted Platform Module (TPM) is all about cryptographic security; the RSA is unbreakable and has constraints concerning data lifetime – data that it protects has some value. So what is all this non-RSA secret stuff about? Well, RSA is a very strong cryptosystem, but as in handling cryptographic keying material, the time and place that a TPM message is transmitted can also be a very important data management issue. The idea of the *rolling nonce* is a basic building block in the protection against a well-known attack, the "man-in-the-middle attack". For example, let's say that there is an "eavesdropper" recording all of the Low Pin Count (LPC) or System Management Bus (SMBus) traffic, regarding TPM input and output command messaging, in an effort to gain the personal computer (PC) resources governed by the TPM. Without some kind of time-in-place management, the TPM would be open to attack from successful command replay and might be manipulated into allowing unwarranted access to system platform resources.

Successful RSA cryptographic protection depends on the strength of the key management facilities as much as TPM command execution depends on placing command execution within the context of a contiguous command sequence. Hence the purpose of the rolling nonce is to provide a mechanism in which to gauge the sequence of TPM command execution and to prohibit command execution outside of this defined sequence executed by the TPM. For that matter, to prohibiting a phony command output to be substituted for an authentic TPM command output inside of the TPM command sequence.

So why rolling nonces? Why not a straightforward protection scheme such as rolling secrets? Well, the TPM could make you change your secret, which is associated with the RSA key, to perform the signing operation each time it is successfully executed. This would indeed stop the TPM command replay since the secret would be different every time the RSA signing operation is invoked; but, is this really a good answer? The short answer is *no* because of the means that are employed to store the keying material or entity secrets inside the TPM, non-volatile memory. Most non-volatile memory has write-cycle limitations regarding the number of times data can be written successfully and since changing the secret based on messages, as opposed to key management operations, could cause the non-volatile memory to fail in a short period of time. So, what do we do to stop TPM message replays from happening? The short answer is the nonce.

OK, what the heck is a nonce? A *nonce*, in Trusted Computing Group (TCG) talk, is simply a 20-byte random value – simple as that. Give me a random number generator, produce a 20-byte random number, and you have a TPM nonce. The lifecycle of the nonce, for the purpose of anti-replay attacks, is one message exchange between the TSG Software Stack (TSS) and the TPM. When a new TPM input/output message block is needed, new nonce pairs are produced and the old nonce pair is null and void. Furthermore, as we will see in the next chapter, nonce pairs are significant to the TPM command authorization protocol. Let's look specifically at the two most basic nonce pairs, how they relate to the TPM command messages, and why they protect the TPM against command replay attacks.

If you skim through the TCG Version 1.1b Main Specification and look at any authorized TPM command, you will see two parameters identified within the authorization block, the nonceEven and nonceOdd. The nonceEven is associated with the TPM output commands and the nonceOdd is associated with the TPM input commands. At first, this may seem like a dyslexic fit on the part of the TPM Specification, until you understand when the first nonce in the chain of rolling nonces is created. You will see that the TPM produces the first nonce by successful execution of the TPM_Object Independent Authorization Protocol (OIAP) command. The TPM_OIAP, along with the other associated TPM authorized commands, will be detailed in Chapter 12, but the fundamental principle is that the nonceEven is produced by the TPM and conveyed through the command output message block. On the other hand, the nonceOdd is produced by the TSS or embedded security stack and is conveyed to the TPM within the command input message block. When the input command is successfully executed within the TPM, the output block, associated with the input message command, will contain a new nonceEven produced by the TPM. Hence the term rolling nonce, each new input message block contains a new nonceOdd produced by the TSS and each successful resulting output message block contains a new nonceEven produced by the TPM.

In addition to generating the nonce's, both entities – the TPM and the external secure stack – have knowledge of both nonce's prior to calculating and verifying the command input and output authorization blocks. Remember, the TPM starts the cycle off via the successful execution of the TPM_OIAP command and shares the internal nonceEven value contained in the command message output block. The TPM gains knowledge of the nonceOdd during the next authorized input message block transaction. This knowledge allows both entities to track the command execution sequence across the LPC or SMBus communication pipe and protect against command replay attacks.

Now that each "entity", the TPM and the TSS, has knowledge of both the nonceEven and nonceOdd, both sides can calculate the authorization digest so that an authorized command can be successfully executed within the TPM. Note there is a lot more to cover on the exact TPM command authorization protocol, which is covered in detail in the next chapter. The point of this discussion is to inform readers about the idea of rolling nonces and how they are produced, used, and rolled. Speaking of rolled, how are these nonce's rolled so to speak? Well we said the TPM produces the nonceEven on the output of the TPM_OIAP command, right? Do you think this is the only command that produces a nonceEven in its output? Well the short answer is *no*. All authorized TPM commands produce a new nonceEven on their output and all authorized TPM input commands require a new nonceOdd within their input command blocks. Hence the idea of the rolling nonce, every new TPM authorized command requires a new nonceOdd and in return, the TPM produces a new nonceEven. Figure 8.1 shows an example nonce chain produced by executing anonymous TPM commands that are chained together.

The most important aspect to remember concerning TPM nonces is that the TPM generates the first nonceEven and is presented at the TPM output message block. In addition, the nonceOdd is produced by the TSS, embedded stack, or development application and supplied to the TPM via the TPM input command message block.

So far the discussion has revolved around the protection architecture based on rolling nonces and not how this protection architecture actually prevents

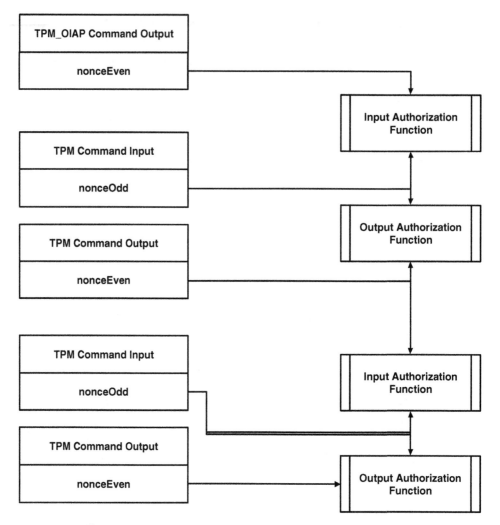

Figure 8.1 Rolling nonces across anonymous TPM commands

command replay attacks against the TPM. For the purpose of this discussion, we will assume that a TPM_Sign command has been previously executed and a command replay attack is about to be performed on the TPM using stale TPM_Sign input vectors. The first item of interest is exactly how the nonceEven and nonceOdd relate to the command authorization protocol and why stale nonces will fail the command authorization. Let's look at an overview of how the nonceEven and nonceOdd are related to the TPM command authorization; in other words, how the TPM command authorization digest calculation leverages the nonceEven and nonceOdd.

First, let's consider the algorithm used to calculate the command authorization digest. That digest comprises a 20-byte Hash-based Message Authentication Code (HMAC) result, which is calculated from a message block that contains the nonceEven and nonceOdd values, along with other pertinent data that is examined in the next chapter. The authorization HMAC is performed within the TPM and the corresponding TSS, with each entity making use of the most recent nonceEven and nonceOdd values stored within their respective boundaries. This sequence of nonce sharing between the TPM and secure software stack allows both entities to autonomously calculate the exact same authorization digest with regard to the "rolling-nonce" dependency. This in effect produces a TPM command authorization within the context of the TPM command sequence of execution. So, if a command replay is attempted by interjecting a command input message that is out of sequence relative to the rolling nonces, the authorization digest check will fail and the command will not be allowed to execute. The reason for this has to due to the fact that the TPM has knowledge of the current nonceEven and nonceOdd and the stale command input block reflects a previous nonce pair that is no longer valid. Since the attacker does not have knowledge of the command entity(s) secret(s), the command authorization digest cannot be calculated taking into account the current nonceEven and nonceOdd values.

Notice that the term *dependency* is used concerning nonce usage regarding the command authorization digest. The fact of the matter is that the nonce values are used to protect the TPM from command message replays and are not directly responsible for command authorization, only part of it. The nonces in the authorization block provide a sequence stamp that protects against reply attacks when combined with the TPM's authorization protocol. In fact any entity can request a nonceEven by executing the TPM_OIAP command, producing a nonceEven via command execution. The same holds true with regard to the nonceOdd value – anyone can generate a random number. All of this nonce generation will only provide the ability to sequence a TPM command event and not authorize such an event without knowledge of the entity's Usage secret.

This is somewhat similar to the analogy that someone can say when they trying to log into a computer, but can not gain access because he or she does not know the user ID and password. Another analogy is when someone trys to use a password that may be correct but is out of context with a password lifetime defined by a timestamp that has expired. The TPM rolling nonces provide a metric concerning relative command sequencing that is used in conjunction with authorization digests that depend on the knowledge of some secret.

Let's take a look at some examples, both successful and unsuccessful regarding the use of TPM rolling nonces. First, a successful command execution – tracking the nonce values and command sequencing – is depicted in Figure 8.2. This example only details the rolling nonces and leaves the other command

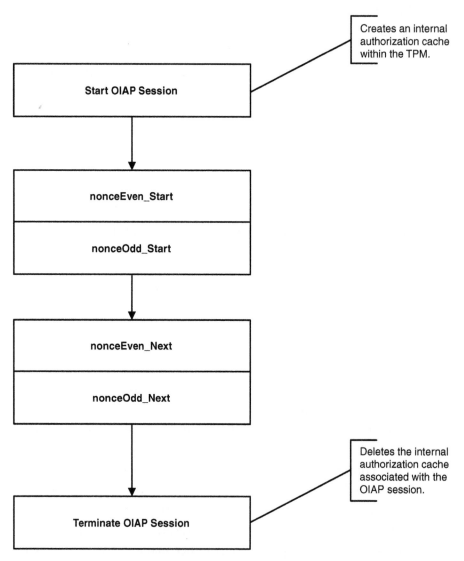

Figure 8.2 Successful nonce sequencing

parameters undefined; it focuses solely on the relationship between the nonce Even, nonceOdd, and command authorization relative to rolling nonces.

The TPM_OIAP command is initially executed, allowing the creation of the first nonceEven and supplying this value to the security stack within TPM_OIAP command output message block. At this juncture, the security stack digests the output message block and stores the nonceEven within its boundaries; so at this point, both the TPM and the security stack have knowledge of the nonceEven. Next, for argument sake, the security stack wishes to perform a TPM_Sign command, but first it must generate a nonceOdd vector and insert this value into the proper place within the TPM_Sign input message authorization block. The TPM_Sign command is sent to the TPM and the nonceOdd is recorded within the TPM boundary, so now the secure stack and the TPM have knowledge of the nonceOdd. Before the command is executed internally, the TPM calculates the command input authorization digest, in which the nonceEven and nonceOdd play a part. The TPM compares the calculated digest to the authorization digest supplied by the secure stack within the command input authorization block. In this case, the nonceEven, nonceOdd, and authorization digest are correct and the TPM allows the signature command to begin executing.

After the signature has been created, the TPM produces a command output message block containing an output authorization block. The secure stack receives the output message block, records the new nonceEven vector that the TPM produces, and calculates an output message authorization digest. The secure stack now compares the authorization digest to the authorization digest produced by the TPM; in this case, the operation passes with the previous nonceOdd and new nonceEven vectors. If the secure stack were to request the TPM to perform another authorized command, a new nonceOdd value would be produced within the secure stack, and the nonceEven value obtained via the output from the TPM_Sign command would be used during authorization digest calculation. Note that this example has been taken out of context with regard to previous TPM state and available RSA keying material that may or may not be loaded within the TPM for use concerning RSA signature operations. More contiguous command sequence examples are given in later chapters.

Let's look at an example of a command replay attack using a previous command input that is out of sequence with regard to the rolling-nonce protocol as shown in Figure 8.3. For this example the TPM_OIAP command is

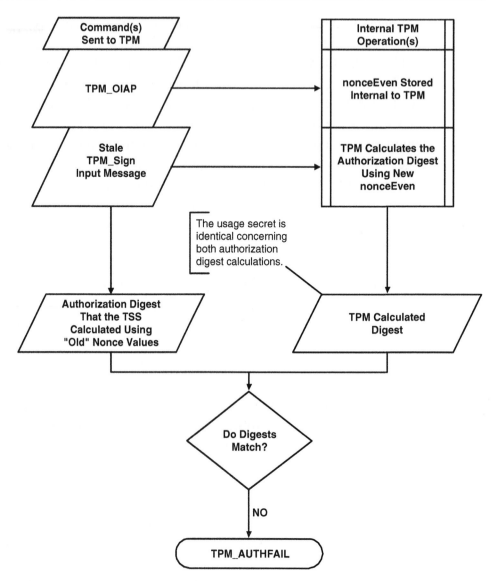

Figure 8.3 Input command replay attack

executed once again, but the TPM_Sign command input block is replayed from the previous example. Right off the bat, the observation can be made that the nonceEven vector produced by the example TPM_OIAP execution will not be the same as the previous example. In addition, the value of the nonceOdd produced by the secure stack is pointless since the command authorization will fail due to the discrepancy between the nonceEven values.

Here the completion of the square is in order. The TPM has recorded the new nonceEven associated with the TPM_OIAP execution and the stale TPM_Sign command message block is sent to the TPM for execution. This time the TPM calculates the authorization digest using the nonceEven produced by the example TPM_OIAP command and the nonceOdd, reused, sent in via the stale replayed command. The authorization digests will fail to match; the stale command authorization digest depends on an incorrect nonceEven value and the TPM will respond to a TCG_AUTHFAIL. Then, you lose.

One other example is in order concerning a replay of a TPM command message output against the secure stack in an attempt to trick the stack into thinking that a valid TPM result was received. Let's look at the depiction of a command replay attack using a previous command output that is out of sequence regarding the rolling-nonce protocol, as shown in Figure 8.4.

In this example, a "man-in-the-middle" or "entity-in-the-middle" is going to try to convince the secure stack that the TPM successfully completed a TPM_Sign command by replaying the previous example's TPM output message block. The TPM_OIAP is once again issued and the TPM_Sign input block is created and sent, but the TPM responds to this command with a TCG_AUTHFAIL. Now suppose that an entity-in-the-middle does not want the secure stack to realize that the TPM_Sign command failed and it tries to replay with a passing TPM_Sign output message block. In this case, there is a double whammy – neither of the nonce values are correct within this command sequence with reference to the output message authorization digest calculation. The secure stack will have produced a new nonceOdd, regarding the failing TPM_Sign input command message, and the nonceEven value given to the secure stack via the replayed output message will be outside of the sequence of rolling nonces. So, when the secure stack calculates the authorization digest and compares it to the TPM's stale calculation, the secure stack will fail the comparison and the output message will be tagged as invalid.

In a basic sense, the idea of the nonce concerns a 20-byte random value, which earmarks a unique place in time regarding the execution of sequential TPM commands. By themselves, the nonceEven and nonceOdd are not very interesting, from a security point of view; but when combined with command authorization, they provide a very formidable barrier against TPM command replay attacks. The next chapter outlines the idea concerning the

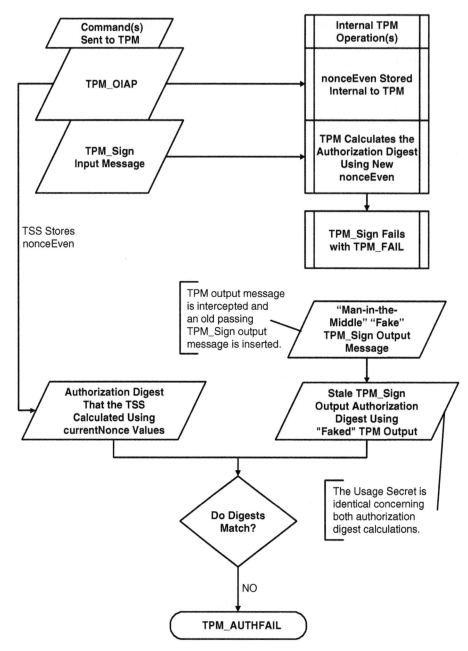

Figure 8.4 Command output replay attack

TPM authorization protocol for both input and output command message blocks. Within this authorization protocol description, the nonceEven and nonceOdd will be further defined with regard to their dynamic use in protecting the TPM from malicious attacks.

9

Command Authorization, Typical

9.1   TPM Authorization Overview

The first thing that must be said concerning Trusted Platform Module (TPM) command authorization is *be very careful.* Do not take this warning lightly. The authorization of TPM commands carry a much heavier penalty than failed execution attempts regarding specific functional commands. For example, the TPM_LoadKey, a TPM function that loads an RSA key pair within the TPM, will tolerate its parameter(s) being modified while integrating this function's successful command message execution during application development. This means that if the TPM_LoadKey fails relative to that function's specific purpose, you can modify the input parameters in an attempt to correct the error and execute the command successfully. This process can be repeated until the development of application code successfully produces the correct parameter block that succeeds in loading a key into the TPM.

I say again, parameters that deal *with specific command execution values*, not the message parameters or operations that support the command authorization. So, instead of hosing up a command-specific parameter, you have an incorrect parameter or calculation regarding the command authorization digest, and you execute based on this type of error. The TPM will fail the input command block with a TCG_AUTHFAIL and, in addition, will keep track of the failed authorization attempts, this attempt being one of them. After a defined number of attempts, regarding command authorization failure, the TPM will protect itself and penalize you in a manner defined by the TPM vendor. This penalty will be severe and will cripple the TPM command development by limiting the command set that can be successfully executed by the TPM.

Some of you may be thinking that this is a radical protection scheme and is completely over the top regarding penalty severity. The reason this is such a severe penalty case has to do with another well-known cryptographic attack protocol, a denial-of-service attack. This book is not intended to define the fine details of cryptographic attacks, but this type of attack boils down to getting information from failed system responses then modifying the attack based on the information divulged by each recurring system failure response. The main point that I am making concerns the TPM command integration within application development and the tendency to "dial-in" successful command compilation. This *dial-in* method will buy you a swift retaliation from the TPM security protection mechanism regardless of the motive behind the attack. This chapter's intent is to define the proper TPM command authorization parameters and authorization digest computation in an effort to avoid invoking the TPM denial-of-service attack protection mechanism.

9.2   The TPM Authorization Input/Output Block(s)

The first step in understanding how the TPM calculates the correct authorization digest is to examine the method of supplying the TPM with parameters that this type of calculation depends on. With this, the authorization block(s) will be defined regarding the authorization parameters as shown in Figure 9.1.

First, the TPM input authorization block attached to the incoming message is examined – TPM's point of view – in the case of a input command authorization requirement. The input authorization block consists of four parameters: the authorization handle, nonceOdd, continue authorization session, and authorization digest. If, per chance, you are looking at Version 1.1b of the TCG Main Specification and you are just by coincidence looking at an authorized command, you might say, "There is a fifth parameter that you forgot to include." Well, you're right and you're wrong. Yes there is a fifth parameter and yes this parameter is an integral piece of data regarding the TPM command authorization digest calculation, but it is not an input parameter.

Look at the specification and notice the first two columns are labeled parameter number (#) and parameter size (SZ). Now look at the nonceEven parameter listed within the authorization block boundary of the input command message definition; hey, "There is no parameter number listed for this entry". That's right, the nonceEven is held within the TPM, as described in the previous chapter. This parameter is listed to aid in viewing the contiguous order of data that will eventually be operated on to produce the authorization digest that is held within the last parameter of the input authorization block.

Authorization Handle	4 bytes	AHB1	AHB2	AHB3	AHB4	
nonceOdd	20 bytes	NOB1	NOB2	NOB3	NOB4	NOB5
		NOB6	NOB7	NOB8	NOB9	NOB10
		NOB11	NOB12	NOB13	NOB14	NOB15
		NOB16	NOB17	NOB18	NOB19	NOB20
Continue Auth Session	1 byte	CAB				
Authorization Digest	20 bytes	ADB1	ADB2	ADB3	ADB4	ADB5
		ADB6	ADB7	ADB8	ADB9	ADB10
		ADB11	ADB12	ADB13	ADB14	ADB15
		ADB16	ADB17	ADB18	ADB19	ADB20

nonceEven	20 bytes	NEB1	NEB2	NEB3	NEB4	NEB5
		NEB6	NEB7	NEB8	NEB9	NEB10
		NEB11	NEB12	NEB13	NEB14	NEB15
		NEB16	NEB17	NEB18	NEB19	NEB20
Continue Auth Session	1 byte	CAB				
Authorization Digest	20 bytes	ADB1	ADB2	ADB3	ADB4	ADB5
		ADB6	ADB7	ADB8	ADB9	ADB10
		ADB11	ADB12	ADB13	ADB14	ADB15
		ADB16	ADB17	ADB18	ADB19	ADB20

Figure 9.1 TPM input and output authorization blocks

In summary, regarding the input authorization block, there are four parameters: the authorization handle (4 bytes), the nonceOdd (20 bytes), the continue authorization session (1 byte), and the authorization digest (20 bytes) for a total of 45 bytes.

Next, the output authorization block, which is attached to a successfully executed authorized TPM command's output message. Looking at Figure 9.1, it can be seen that there are similarities and differences when juxtaposing the input and output authorization blocks. First of all, the output authorization block contains the following: nonceEven, continue authorization session, and authorization digest. Note that the TPM produces a new nonceEven inside the output authorization block, just as defined within the previous chapter, and the nonceOdd is known by the entity that called the TPM function supplying the nonceOdd to the TPM. One other thing, the output authorization message block does not contain an authorization handle. Therefore, the output authorization block contains a nonceEven (20 bytes), continue authorization session (1 byte), and the authorization digest (20 bytes). That's it folks, there are two types of authorization blocks – input and output – and these blocks will remain in these defined formats throughout the entire Trusted Computing Group (TCG) Specification. Notice I didn't mention the number of attached authorization blocks; we will see these on the flip side.

9.3   Types of Command Authorization(s)

Before diving into the command authorization calculation, it would be wise to define authorization levels that the TPM subscribes to and enforces. The quick answer is three, yes three, I can feel my ears burning from a whisper saying, "Two, there is only two". Well, there are only two instances of a command authorization block being attached to command message blocks, but there are three: no authorization, one authorization, and two authorizations. The TPM must make a decision regarding the command-level authorization that needs to be performed and this includes the decision that the command does not require any command authorization.

The first parameter contained within every valid TPM command message block holds a value that represents the command message block's authorization level. Notice that I said "the command's MESSAGE block" and not the COMMAND's authorization level. This is a huge distinction and will become clear as this chapter's detail unfolds; but for now, understand that the tag parameter *within* the command message block pertains to the message block, not the command referenced within the message block. Figure 9.2 shows the possible input message tag values and the following explains these values.

There are three possible input message block tag parameter values that are correct. The first value is 0xC1, which tells the TPM that the message being sent to it contains no attached authorization block. This does not mean that the TPM will not verify that the command reference in the message block can

Non-authorized Command Tag	0xC1
Single Authorized Command Tag	0xC2
Dual Authorized Command Tag	0xC3

Figure 9.2 Input tag parameter values

be legally executed as a non-authorized command, just that the input message block has no authorization information attached. In addition, the tag value 0xC2 tells the TPM that the message being sent to it contains one authorization block and 0xC3 proclaims that two authorization blocks are attached to the message. These are the only valid input message block tag values that are legal and the TPM will error with a return result of TCG_BADTAG if one of these three values is not present. We will discuss the reason for differentiating the message authorization level from the command authorization level later in this chapter and the next chapter, which focuses on command authorization rule exceptions.

Non-authorized Command Tag	0xC4
Single Authorized Command Tag	0xC5
Dual Authorized Command Tag	0xC6

Figure 9.3 Output tag parameter values

The other half of the equation involves the output message block tag parameter, which also has three possible values. Figure 9.3 defines these values. Again, there are three different levels concerning output message block authorization and, like the input tag, these authorization levels pertain to the command message authorization level, not the command that has successfully executed. The first value is 0xC4 and defines a non-authorized message output. The values 0xC5 and 0xC6 define single and double authorization levels, respectively. The three values defined are the only legal output message tag response codes allowed; if some other value is placed in this parameter slot within a TPM output message, you have a serious – like get rid of the TPM – problem.

In conclusion, the tag parameter defines the level of authorization pertaining to the message and not the command specifically. To give a little taste, the TPM_LoadKey, as defined by Version 1.1b of the TCG Main Specification, is a signal authorized command, where *authorization* is relative to the parent key used to wrap the private information regarding the key about to be loaded into the TPM. If the parent key was loaded into the TPM and defined as "no authorization required" to use this key resource, then the TPM_LoadKey message block, in which the Key Handle references the parent key described before, will contain a tag value defining a non-authorized command message block. Hence, the tag parameter defines the level of authorization for the message block not the command; in this case, the Key Handle pointing to the parent key will gauge the ability of the TPM to execute the TPM_LoadKey command as a non-authorized command. If this scenario seems confusing, please be patient; the rest of the chapter along with the next will make this authorization ability clearer.

9.4   Object Independent Authorization Protocol

Behind the scenes I have been debating, with myself, concerning the authorization principal to unfold next. The internal war was fought and the story of where nonce values come from seems in order. You already understand that the TPM generates the nonceEven and conveys this nonce at the time of output message transmission, but we haven't yet described the detailed method of instructing the TPM to produce the first nonceEven. Hence the TPM_OIAP command stands for Object Independent Authorization Protocol (OIAP). OK, what the heck is an *object* in the land of TPM.

Objects or, better, *entities* that refer to things such as the Owner of the TPM, the Storage Root Key (SRK), and User keys. When we say *object independent*, this implies that the authorization protocol is not dependent on or tied to a specific object. The next topic of discussion centers on the Object Specific Authorization Protocol (OSAP) that does tie the authorization to a specific entity, but first the OIAP.

Figure 9.4 shows the input message values that make up the TPM_OIAP command concerning the input message block. Notice that the input tag parameter defines the command as non-authorized. This is so because the TPM_OIAP not only produces the first nonceEven but also invokes the TPM to create an internal authorization cache that can be viewed as an internal authorization scratch pad area. This scratch pad area within the TPM stores authorization data relative to an authorization handle, which is produced during the output of the TPM_OIAP command. The authorization cache

Authorization Tag	2 bytes	0x00	0xC1		
Parameter Size	4 bytes	0x00	0x00	0x00	ox0A
Command Ordinal	4 bytes	0x00	0x00	0x00	0x0A

Figure 9.4 TPM_OIAP input message block

will persist until an authorized command referencing the authorization handle, produced by a TPM_OIAP command, fails or the cache is explicitly destroyed by invoking the command TPM_Terminate_Handle. With that said, let's look at the TPM_OIAP parameter list. First is the infamous tag parameter that we all know and love, followed by a parameter size, and finally the command ordinal, in this case 0x10 (TPM_ORD_OIAP). That's it. This is a good reference for one of the simplest input command messages to create.

The TPM_OIAP output message block contains a tag parameter, which declares the output message to be non-authorized. The tag is followed by a message parameter size and a return code stating the execution results, passed or some failure code. If the command fails, for example TCG_RESOURCES, these will be the only parameters contained within the command output message block. For your edification, TCG_RESOURCES declares that the TPM has consumed all its resources and is unable to create another authorization cache internally. In addition, you can only create a certain number of authorization sessions within the Atmel TPM. If the TPM_OIAP command returns a passing result, ReturnCode = 0x00000000, there will be two additional parameters within the output message: an authorization handle (4 bytes) and a nonceEven (20 bytes). A note to embedded developers who are developing their own security stack: *Do not lose this information*; retain the authorization handle and nonceEven parameters because you will need these values at a later time, I promise.

Figure 9.5 defines a structure that can be overlaid onto a buffer that contains the TPM_OIAP command response. Remember, the authorization handle along with the nonceEven will not be produced within the output of this command if there was an execution failure.

In summary, the execution of the TPM_OIAP command will return, in addition to the tag, parameter size and return code, an authorization handle, and a nonceEven parameter. Also, the command will instruct the TPM to set aside or consume one of its two authorization cache areas in anticipation

Authorization Tag	2 bytes	0x00	0xC4			
Parameter Size	4 bytes	0x00	0x00	0x00	0x22	
Return Code	4 bytes	0x00	0x00	0x00	0x00	
Authorization Handle	4 bytes	0x00	0x00	0x00	0x00	
nonceEven	20 bytes	0xA5	0xA5	0xA5	0xA5	0xA5
		0xA5	0xA5	0xA5	0xA5	0xA5
		0xA5	0xA5	0xA5	0xA5	0xA5
		0xA5	0xA5	0xA5	0xA5	0xA5

Figure 9.5 TPM_OIAP output message block

of an impending authorized command message. This authorization protocol is independent of any TPM entity.

9.5 Calculating the Authorization Digest

Now that we have executed the TPM_OIAP command, we have all the data we need to produce our first authorization digest; note that we will only be calculating a digest based on command message fragments. In later chapters, there will be plenty of full command input message creation along with message output analysis. For now, let's look at a generic template regarding datum used as input(s) to the function that will create the 20-byte authorization digest.

Figure 9.6 overlays the various command input message parameters that will be extracted and processed during the creation of intermediate and final digest values. This example refers to the TPM_Sign command input and output message blocks, bringing a real-world flavor to the discussion. First, as seen in Figure 9.6, the input message block contains 10 separate parameters including the three "header" parameters, the tag, parameter size, and the command ordinal. In addition, there is a Key Handle, area to sign size, and area to sign. The authorization block – this is a single authorization command – is attached to the input message block.

Moving on, the generic description concerning the final step in calculating the authorization digest is to the HMAC (Hash-based Message Authentication Code) four parameters; referencing the TCG Specification, they are 1H1, 2H1, 3H1, and 4H1. If you look to the authorization block within the TPM_Sign command description, you will see the last nonceEven, nonceOdd, and continue authorization session tagged as 2H1, 3H1, and 4H1, respectively. Also

Authorization Tag	2 bytes	0x00	0xC2			
Parameter Size	4 bytes	0x00	0x00	0x00	0x53	
Command Ordinal	4 bytes	0x00	0x00	0x00	0x3C	
TCG Key Handle	4 bytes	0x00	0x00	0x00	0x02	
Area to Sign Size	4 bytes	0x00	0x00	0x00	0x14	
Area to Sign	20 bytes	0xB3	0xD5	0xCB	0x12	0x73
		0x8B	0xB6	0xF9	0x21	0xA3
		0xDA	0x42	0xE0	0x18	0xD1
		0x43	0xFA	0x29	0x7C	0xA6
Authorization Handle	4 bytes	0x00	0x00	0x00	0x0A	0x00
nonceOdd	20 bytes	0xB9	0x73	0x05	0xFA	0xDB
		0xE3	0x4D	0xC5	0x46	0x65
		0x10	0x00	0x0A	0x55	0x04
		0x2E	0x3F	0xEA	0xBF	0x27
Continue Auth Session	1 bytes	0x01				
Authorization Digest	20 bytes	0x26	0x7E	0xCA	0x16	0xA1
		0x4D	0x36	0xE6	0x72	0x2E
		0xAA	0x7F	0x7B	0x53	0x4A
		0xB3	0xCE	0x8B	0x2A	0xAA

nonceEven	20 bytes	0xA5	0xA5	0xA5	0xA5	0xA5
		0xA5	0xA5	0xA5	0xA5	0xA5
		0xA5	0xA5	0xA5	0xA5	0xA5
		0xA5	0xA5	0xA5	0xA5	0xA5

Input data to the SHA-1 hash function; output digest is the 1H1 parameter.

1H2 parameter, nonceEven, which is not part of the input message block.

1H3 parameter, the nonceOdd digest.

1H4 parameter, the continue Auth Session parameter.

Figure 9.6 TPM_Sign input message block

realize that the last nonceEven is not part of the input message that will be sent to the TPM; this value was previously recorded after the last successful authorized TPM command, in this case, the TPM_OIAP.

OK, now I hear the question shouting out from everyone following along so far: "Where is the 1H1 parameter defined?" Good question. Notice in the same column where 1H values are represented, above these values in the command input parameter block, there are other parameters tagged as 1s, 2s, and 3s within the body of the TPM_Sign command. These values point to the input parameters that are to be SHA-1 (Secure Hash Algorithm) hashed and the output of this SHA-1 hash will contain the 1H1 parameter, meaning all four HMAC parameters are defined.

Figure 9.7 defines this flow, with regard to the parameter definition, concerning the four input values that will be HMACed to create the authorization digest.

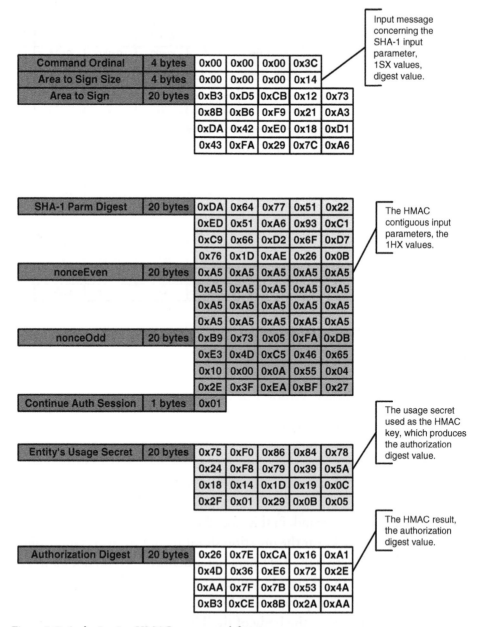

Figure 9.7 Authorization HMAC parameter definition

```
AuthDigest = HMAC [SHA1(InputParms)
                  |nonceEven|nonceOdd|contAuthSess]_UsageSecret
```

This can be confusing at first since the TCG Version 1.1b Main Specification doesn't explicitly define the 1H1 value representation when looking at command parameter definitions. This issue is discussed with regard to the TCG Specification in the section that explains authorization.

All command authorization(s) calculate(s) the 1H1 input parameter using this same rule, perform a SHA-1 hash on the input command message (the command ordinal), and any payload associated with the command. Payload refers to parameters specific to the functionality of the command about to be executed. In the case of the TPM_Sign command, these parameters are the area to sign size and the area to sign. Also, when calculating the SHA-1 digest, the parameters to hash must be in contiguous order. For example, if you remove the Key Handle from the input message block and push the area to sign size and area to sign down 4 bytes, you would produce a continuous block of input parameters that can now be hashed.

The reason I describe this flow is that all commands that have Key Handles defined within in their input message blocks can used this method to calculate the first HMAC parameter. Once the SHA-1 hash is complete simply push the parameters up 4 bytes and place the Key Handle back into its rightful position within the input message block. Note that this procedure works when two Key Handles are present within the command message block; on alteration, move the command payload down 8 bytes instead of 4 bytes. One other thing: any parameters that exist between the ordinal and the authorization block(s), if there is no Key Handle, is considered to be the command payload area.

So, we have all of the HMAC parameters defined; one problem, "Which key do we use for the HMAC?" Remember, an HMAC function is simply an SHA-1 operation that accepts a 20-byte key value as an input in addition to the message to be hashed. Well the answer to the question resides with the entity of that the Key Handle – the same one that was removed then replaced between SHA-1 operations performed on the input message block. This Key Handle references a key that resides within the TPM; note that this key needs to have been loaded within the TPM prior to performing a signature operation. Associated with this key is a Usage Secret, hence the HMAC key.

Don't snap a brain cell worrying about where that secret comes from or how it got placed in the TPM; the answer awaits you in Chapter 15, which explains the TPM_LoadKey command. Just realize that the TPM keeps track of a Usage Secret associated with a key it loaded after a successful TPM_LoadKey command. Also, realize that the TCGSoftware Stack (TSS) must have knowledge of this secret since it produced the TPM_LoadKey input command message and had to supply this information for the command to be successful.

Once the HMAC key inferred by the Key Handle parameter is known, it is a simple matter to perform the HMAC operation to acquire the authorization digest for the authorized command. Figure 9.8 shows a block diagram that outlines the input and output parameters concerning the HMAC function.

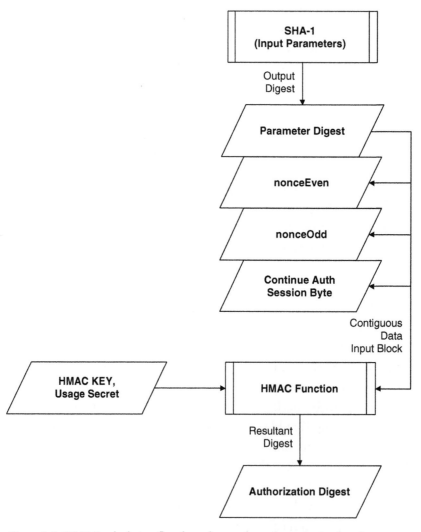

Figure 9.8 HMAC calculation flowchart that produces the authorization digest

Simply, execute the HMAC function on the input parameters using the entity Usage secret referred by the Key Handle and "Bam" (I know someone famous says that), you have produced the authorization digest. Now you can place this digest into the authorization block of the command input message and you're good to go. One other thing: there is no difference between the TPM and TSS concerning the calculation of the authorization digest – the TPM performs the same operation internally as the TSS did to produce the digest and authorize the command.

9.6   Object Specific Authorization Protocol

Now we come to the interesting part concerning authorization sessions. In the previous section, the discussion focused around independent authorization sessions or authorization sessions that are not tied to any TPM entity. Now someone might say that the Key Handle tied the authorization calculation to an entity, in this case a user key, but it was the specific command, which reference the entity. No parameters within the authorization block attached to the command input message or output message referenced any TPM entity. Therefore, the same OIAP session could perform a signature operation with one key entity and turn around and perform a load key operation referencing another key entity, hence the term *independent*. The OSAP does tie into a single entity and will implicitly reference that entity within the authorization block via a Shared Secret.

First, let's take a look at the input message block that makes up the entire TPM_OSAP command, as shown in Figure 9.9. The initial observation that can be made is that the OSAP command has more input parameters than the OIAP command, the tag, parameter size, and command ordinal. The OSAP

Authorization Tag	2 bytes	0x00	0xC1			
Parameter Size	4 bytes	0x00	0x00	0x00	0x24	
Command Ordinal	4 bytes	0x00	0x00	0x00	0x0B	
Entity Type	2 bytes	0x00	0x05			
Entity Value	4 bytes	0x00	0x00	0x00	0x02	
nonceOSAPOdd	4 bytes	0xB9	0x73	0x05	0xFA	0xDB
		0xE3	0x4D	0xC5	0x46	0x65
		0x10	0x00	0x0A	0x55	0x04
		0x2E	0x3F	0xEA	0xBF	0x27

Figure 9.9 TPM_OSAP input message block

command adds three new parameters: the entity type, entity value, and nonceOddOSAP (not another nonce). Yes, another nonce and remember this: the nonceOdd *is not* the same value as the nonceOddOSAP. Do not confuse these two values, surprisingly this is a very common first mistake and I have fielded many calls from someone who declared that the authorization calculation based on the OSAP was wrong only to discover that the nonce used to calculate the Shared Secret was the nonceOdd and not the nonceOddOSAP. Before the details concerning the Shared Secret are revealed, let's look at the output message block for the TPM_OSAP after a successful execution.

Authorization Tag	2 bytes	0x00	0xC4			
Parameter Size	4 bytes	0x00	0x00	0x00	0x36	
Return Code	4 bytes	0x00	0x00	0x00	0x00	
Authorization Handle	4 bytes	0x00	0x00	0x00	0x00	
nonceEven	20 bytes	0xA5	0xA5	0xA5	0xA5	0xA5
		0xA5	0xA5	0xA5	0xA5	0xA5
		0xA5	0xA5	0xA5	0xA5	0xA5
		0xA5	0xA5	0xA5	0xA5	0xA5
nonceEvenOSAP	20 bytes	0xA5	0xA5	0xA5	0xA5	0xA5
		0xA5	0xA5	0xA5	0xA5	0xA5
		0xA5	0xA5	0xA5	0xA5	0xA5
		0xA5	0xA5	0xA5	0xA5	0xA5

Figure 9.10 TPM_OSAP output message block

Figure 9.10 describes the TPM_OSAP command output message. Notice that the OSAP command output message contains the exact output parameters relative to the OIAP command output message with one exception: the OSAP output message contains an extra nonce. Attached below the nonceEven parameter is an additional 20 bytes representing the nonceEven OSAP – where there's a nonceOdd there is a nonceEven.

So what does all this imply relative to the OSAP functionality regarding rolling nonces and anti-replay attacks performed on the TPM? Nothing. The OSAP uses the nonceEven and nonceOdd in the exact same way. The nonceEvenOSAP is sent through the OSAP output message to give the TSS ability regarding the Shared Secret calculation, not so different from the information exchanged during command authorization. So, the nonceEven is used in the same manner as defined by the overall command authorization scheme, and the nonceEvenOSAP is related to the calculation of the Shared

Secret. Let's look at the calculation regarding the Shared Secret and how this secret ties the OSAP authorization to a single entity.

The Shared Secret is similar to the usage authorization associated with keying material that binds the use of the key to the user of that key; each value takes the form of an HMAC key during authorization digest calculation. The difference being that the Usage Secret is tied only to an entity and the Shared Secret is tied to an entity and an authorization session created by the OSAP, which implies that the authorization session is bound to an authorized command referencing the use of that particular entity. To see how this bind between the OSAP and entity is created, the Shared Secret calculation must be described.

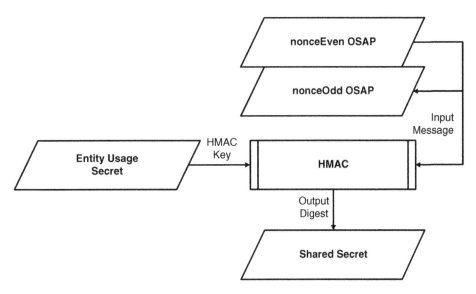

Figure 9.11 OSAP Shared Secret calculation

Figure 9.11 shows this calculation using the entities Usage Secret, the nonceEvenOSAP and the nonceOddOSAP. The calculation of the Shared Secret is simple: HMAC the concatenation of the nonceEvenOSAP and nonce OddOSAP using the entity's Usage Secret as the HMAC key.

$$\text{SharedSecret} = \text{HMAC}(\text{nonceEvenOSAP}|\text{nonceOddOSAP})_{\text{UsageSecret}}$$

This Shared Secret now becomes the equivalent Usage Secret when referencing the calculation of the authorization digest which is used to authorize OSAP-based command input and output messages, using the same algorithm when based on the OIAP authorization protocol. In other words, replace the Shared Secret with the Usage Secret during OSAP-based authorization.

The key piece of information is the HMAC key, used to authorize the OSAP-based command(s), that is derived from the OSAP nonce's and the entity's Usage Secret binding the OSAP to the authorization session and the entity. Figure 9.12 describes the overall relationship between the OSAP, the entity's Usage Secret, and the command authorization.

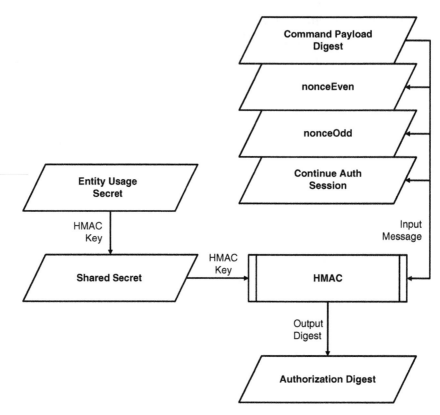

Figure 9.12 Binding of OSAP to entity usage

The last subject concerning OSAP relates to a discussion on Entity Types and Entity Values – the mechanism that is used to attach OSAP session(s) to Entity Objects or Usage Secrets. There exist several types of entities that can reside within the TPM and be referenced by the OSAP session in regard to binding the authorization session to said entity. Figure 9.13 defines the possible entity types that can exist within the TPM.

Notice that the first Entity Type is the TCG_ET_KEYHANDLE and, if selected, the Entity Value would be the associated Key Handle. The next Entity

TCG_ET_KEYHANDLE	0x00	0x01
TCG_ET_OWNER	0x00	0x02
TCG_ET_DATA	0x00	0x03
TCG_ET_SRK	0x00	0x04
TCG_ET_KEY	0x00	0x05

Figure 9.13   TCG Entity Types

Types that are of extreme interest are the TPM Owner-based Entity Types: the TCG_ET_OWNER and the TCG_ET_SRK. The reason owner-based entity type is used to describe these values concerns the nature of the association with the TPM Owner, the TCG_ET_OWNER is an explicit association and the TCG_ET_SRK is an implicit cryptographic association. When referring to these two Entity Types, the associated Key Handle concerning TCG_ET_OWNER and TCG_ET_SRK must be 0x40000001 and 0x40000000, respectively. These are "special" handles that explicitly refer to predefined entities such as the TPM Owner and SRK. The last two Entity Types refer to generic data or keying material, which is a blob of data or generic key data not associated with a Key Handle. Detailed descriptions concerning Entity Types and their use is deferred to specific commands that leverage their definition and use.

The OSAP session must be terminated on specific conditions, as follows:

1. The Entity, which the OSAP session is bound to, is unloaded.
2. The Entity's authorization – Usage Secret – has been modified.
3. The OSAP session was used to execute a TPM_ChangeAuth command.
4. The command being executed, via the OSAP session, fails.

The OSAP session can be explicitly deleted from the TPM and its internal authorization cache freed by the successful execution of the TPM_Terminate_Handle command.

To summarize, the OSAP command links or binds the authorization session to a specific entity and only allows the use of the authorization session to that specific entity. The idea of a Shared Secret allows realization concerning this concept through the use of two nonces: the nonceEvenOSAP and the nonce OddOSAP. Calculation of the Shared Secret is performed via an HMAC on the concatenation of the nonceEvenOSAP and nonceOddOSAP with the HMAC key assigned to the entity's Usage Secret. The OSAP defines the Entity Type and Entity Value to explicitly index the entity to be bound to the authorization session. Any variation of the entity's state – the entity is unloaded or the authorization secret has being modified – means the TPM_ChangeAuth command or command execution failure will result in the termination of the OSAP session.

9.7   Command Authorization Examples, Typical

Now to tie together a sequence of TPM commands that will provide examples outlining the execution beginning with OIAP then OSAP through authorization session termination. Note that these examples rely on the creation of a TPM Owner, and in doing so, rely on the SRK being present within the TPM. For more information concerning TPM ownership, see Chapter about ownership of the TPM. In addition, these examples will leverage the concept; notice I didn't use the term *mode* concerning TPM compliance vectors; see Chapter 12, which explains compliance vectors and their use.

Briefly, compliance vectors allow testing and application development to proceed within an environment of low entropy – no random-based data. This allows the test or development results to follow a line of predictability, but if used could mask problems in key management and other data management realizations. Such problems are normally due to the fact that nonce values, along with other random values, are predictable within the compliance state, and that complacency can lead to issues where these value(s) are used without regard to how they are managed. The moral of this story is: if compliance vectors are used to develop code, clear the TPM and run the application using truly random data that will flush out any data management problems relying on the consistency inherent within the compliance vectors themselves.

Example 1: Load Key and Sign Using an OIAP Session

This first example loads a key into the TPM, after a successful TPM_OIAP has been executed, and signs a message with this key referencing the same OIAP session. First, the OIAP authorization session must be established, so execute the input message command depicted in Figure 9.14. After the successful

Authorization Tag	2 bytes	0x00	0xC1		
Parameter Size	4 bytes	0x00	0x00	0x00	0x0A
Command Ordinal	4 bytes	0x00	0x00	0x00	0x0A

Figure 9.14 Initial command sequence OIAP

Authorization Tag	2 bytes	0x00	0xC4			
Parameter Size	4 bytes	0x00	0x00	0x00	0x22	
Return Code	4 bytes	0x00	0x00	0x00	0x00	
Authorization Handle	4 bytes	0x00	0x00	0x00	0x00	
nonceEven	20 bytes	0xA5	0xA5	0xA5	0xA5	0xA5
		0xA5	0xA5	0xA5	0xA5	0xA5
		0xA5	0xA5	0xA5	0xA5	0xA5
		0xA5	0xA5	0xA5	0xA5	0xA5

Figure 9.15 OIAP output message

execution of this command, an output message will be available to the TSS for consumption, as shown in Figure 9.15.

Referring to the output message of the TPM_OIAP command, notice the nonceEven generated by the TPM; it is a 20-byte fixed value of 0xA5. This is an example pertaining to one of the characteristics concerning compliance vectors, no random data. When the time comes to produce the TPM_LoadKey input message, the fixed value concerning the nonceOdd will be looked at. After consuming the output of the TPM_OIAP command, the TSS, or test application, will record the authorization handle and the nonceEven values for future use.

Now that the OIAP command has successfully executed, the TPM has allocated cache memory for session scratch pad use and has supplied the TSS with an authorization handle, indexing this internal session, and the nonceEven value that will be used in authorization digest calculation. The next step is to create the TPM_LoadKey input message using the newly created OIAP session as the authorization vehicle. First, the input message must be defined with regard to the Load Key command; then, the authorization digest can be calculated and populated within the Load Key command's authorization block. The parent key, used to encrypt private data associated with the key about to be loaded, will be the SRK for this particular example. Figure 9.16 depicts the TPM_LoadKey input command message and shows the input(s) along with the series of calculations concerning the authorization digest that will allow this command to be authorized and executed.

Note that the purpose of this discussion is to outline the creation of the authorization block concerning the TPM_LoadKey command, not the command-specific parameters themselves. With this said, please ignore the specific parameters with regard to the TPM_LoadKey and focus of the SHA-1 hash parameters definition along with the corresponding HMAC. If you are the type that wants all details defined before moving on, now would be a good time to check out Chapter 15, which outlines the TPM_LoadKey specifics.

With that said, let's discuss the specific authorization digest calculation and the resulting authorization message block that will be attached to the

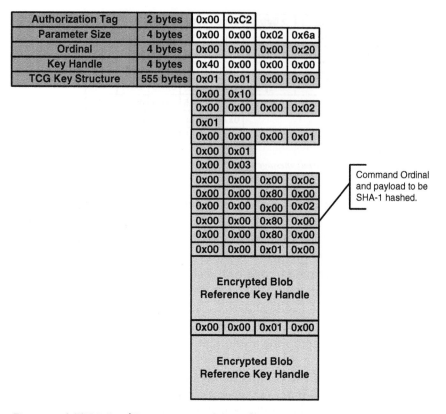

Authorization Tag	2 bytes	0x00	0xC2		
Parameter Size	4 bytes	0x00	0x00	0x02	0x6a
Ordinal	4 bytes	0x00	0x00	0x00	0x20
Key Handle	4 bytes	0x40	0x00	0x00	0x00
TCG Key Structure	555 bytes	0x01	0x01	0x00	0x00
		0x00	0x10		
		0x00	0x00	0x00	0x02
		0x01			
		0x00	0x00	0x00	0x01
		0x00	0x01		
		0x00	0x03		
		0x00	0x00	0x00	0x0c
		0x00	0x00	0x80	0x00
		0x00	0x00	0x00	0x02
		0x00	0x00	0x80	0x00
		0x00	0x00	0x80	0x00
		0x00	0x00	0x01	0x00

Command Ordinal and payload to be SHA-1 hashed.

Encrypted Blob Reference Key Handle

0x00	0x00	0x01	0x00

Encrypted Blob Reference Key Handle

Figure 9.16 TPM_LoadKey input message example

TPM_LoadKey command. First off, Command Ordinal and Command Payload data must be in contiguous blocks. I don't care how you manage that; for example, if you have plenty of memory (all you PCers can relate to this), then you might want to set aside a specific region that you can copy the pertinent data to in lieu of the authorization digest. Otherwise, the rest of us (the poor saps that beg, barrow, and steal every byte that we can get) will have to do the authorization digest calculation in place. This means adjusting the Key Handle – the 4 bytes after the message parameter size – by saving off this value and moving the payload down against the Command Ordinal (see the previous paragraph that discusses this concept).

After the ordinal and command payloads are in contiguous formats, all that is left to do is hash, SHA-1, the data and produce the 20-byte 1H1 parameter. Moving on, in some predefined scratch pad area (sorry, you will have to set aside a cache area to perform the authorization digest calculation), concatenate the 1H1, 1H2, 1H3, and 1H4 parameters. In realistic terms concatenate the parameter digest, the nonceEven, the nonceOdd, and the continue authorization session. When this is done, HMAC blobs this with the SRK usage authorization; we are using compliance vectors so this usage

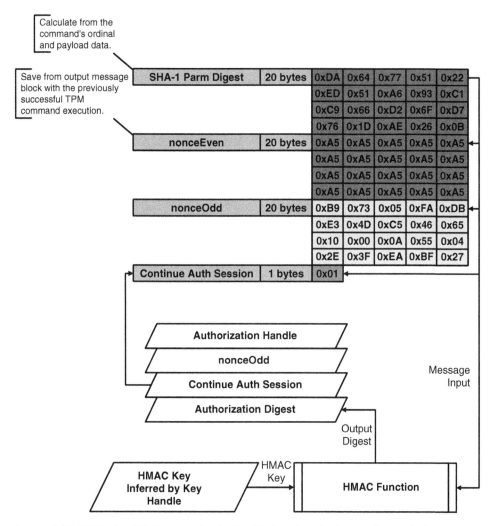

Figure 9.17 TPM_LoadKey input authorization block and authorization digest calculation

authorization will be 20 bytes of 0xA5s. Figure 9.17 defines this calculation based on the input message shown in Figure 9.16.

Now attach the Authorization Message Block to the TPM_LoadKey command message block and presto you have a fully defined TPM_LoadKey input message block ready to send to the TPM for execution. Note that this is not entirely true if you have not defined a TPM Owner and this example's purpose is entirely based on calculating an authorization message block. Later on, a contiguous command sequence will be outlined with the sole purpose of actual command execution.

OK, a little smoke and mirrors, and we have successfully executed the TPM_LoadKey command, which produces the output message block shown in Figure 9.18. The output message needs to be consumed by the TSS, and

Authorization Tag	2 bytes	0x00	0xC2			
Parameter Size	4 bytes	0x00	0x00	0x02	0x6a	
Ordinal	4 bytes	0x00	0x00	0x00	0x20	
Key Handle	4 bytes	0x40	0x00	0x00	0x00	
TCG Key Structure	555 bytes	0x01	0x01	0x00	0x00	
		0x00	0x10			
		0x00	0x00	0x00	0x02	
		0x01				
		0x00	0x00	0x00	0x01	
		0x00	0x01			
		0x00	0x03			
		0x00	0x00	0x00	0x0c	
		0x00	0x00	0x80	0x00	
		0x00	0x00	0x00	0x02	
		0x00	0x00	0x80	0x00	
		0x00	0x00	0x80	0x00	
		0x00	0x00	0x01	0x00	
		Encrypted Blob Reference Key Handle				
		0x00	0x00	0x01	0x00	
		Encrypted Blob Reference Key Handle				
Authorization Handle	4 bytes	0x00	0x00	0x00	0x00	
nonceOdd	20 bytes	0xB9	0x73	0x05	0xFA	0xDB
		0xE3	0x4D	0xC5	0x46	0x65
		0x10	0x00	0x0A	0x55	0x04
		0x2E	0x3F	0xEA	0xBF	0x27
Continue Auth Session	1 bytes	0X01				
Authorization Digest	20 bytes	DB1	DB2	DB3	DB4	DB5
		DB6	DB7	DB8	DB9	DB10
		DB11	DB12	DB13	DB14	DB15
		DB16	DB17	DB18	DB19	DB20

Figure 9.18 TPM_LoadKey output message block

before the TSS can realistically use this data, an authorization digest must be calculated independently and matched to the authorization digest contained in the authorization message block attached to the TPM_LoadKey output message block.

This involves a slightly different protocol with regard to output parameter hashing, which involves a parameter that must survive from the input message block, the ordinal. Putting a finer point on the subject: The TPM_LoadKey output parameter digest is calculated from the concatenation of three parameters – the return code, the ordinal, and the output payload, if any. The ordinal is not contained within the output message block and must persist from the transmission of the input message block through command execution.

After this data is in contiguous form, simply SHA-1 the message to produce the 1H1 parameter. As usual, to complete the authorization digest calculation, concatenate the 1H1, 1H2, 1H3, and 1H4 parameters, and HMAC this message using the SRK Usage Secret. The four 1HX parameters are the typical players: the parameter digest, the nonceEven, the nonceOdd, and the continue authorization session. One concept worth mentioning involves the nonceEven digest. This is *not* the same nonceEven value used during the input command authorization calculation. The TPM has produced a new nonceEven per definition stating that the TPM produce a new nonceEven on all authorized command output(s).

When the authorization digest is independently calculated, the only task that remains is to compare the transmitted authorization digest to the calculated digest. If the two values match, the TPM_LoadKey command output parameters can be consumed and used. Figure 9.19 outlines the independent authorization digest calculation along with the digest comparison.

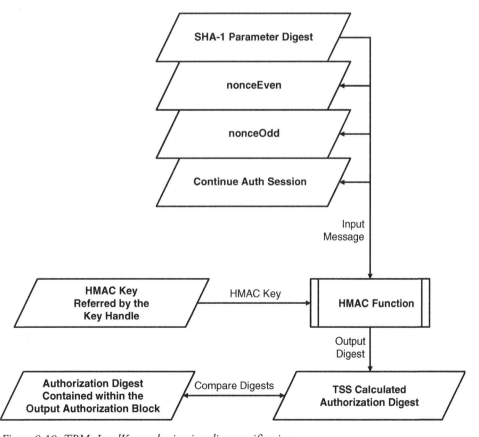

Figure 9.19 TPM_LoadKey authorization digest verification

That's it. After validating the output authorization digest, the TSS can now safely store the TPM Key Handle that points to the newly loaded RSA keying material the TPM and associated specific key properties that it defined when building the TPM_LoadKey input message block. Again, see Chapter 15, devoted to the TPM_LoadKey command for specifics regarding individual command parameters. One more side note: the nonceEven that has been passed along to the TSS via the output message block can be used within another command input message block if a sequential command is to follow. In this example we are going to do just that. Now that, in this case, the signing key has been loaded within the TPM, we will leverage this and sign a SHA-1 digest.

Authorization Tag	2 bytes	0x00	0xC2			
Parameter Size	4 bytes	0x00	0x00	0x00	0x53	
Ordinal	4 bytes	0x00	0x00	0x00	0x3C	
Key Handle	4 bytes	0x00	0x00	0x00	0x01	
Size of Message	4 bytes	0x00	0x00	0x00	0x14	
Message to Sign	20 bytes	0xB3	0xD5	0xCB	0x12	0x73
		0x8B	0xB6	0xF9	0x21	0xA3
		0xDA	0x42	0xE0	0x18	0xD1
		0x43	0xFA	0x29	0x7C	0xA6

Figure 9.20   TPM_Sign input message block

First things first. Let's look at the input command message block regarding the TPM_Sign command as defined in Figure 9.20. This should be no surprise (I am sure hoping this is true) in regard to the "header information". The command payload consists of the following parameters: the Key Handle; the output handle from the TPM_LoadKey we just executed: the area to sign size, in this case it is 20 bytes: a SHA-1 digest; and the area to sign, the digest itself.

Same old procedure – make a contiguous payload region by removing the Key Handle and pushing down the area to sign size and area to sign. SHA-1 hash the payload region and produce the 20-byte digest representing the payload hash, concatenate this value to the nonceEven, nonceOdd, and the continue authorization session parameters. HMAC the 1H1, 1H2, 1H3, and 1H4 parameters using the Usage Secret assigned to the Key Handle – that's right the Key Handle *not* the SRK as used within the TPM_LoadKey command. We are signing with the previously loaded signing key not the SRK, which is an encryption key.

Figure 9.21 defines the calculation. Notice that the TSS *must* produce a new nonceOdd that is in line with the definition that all nonceOdd digests will be produced within authorized TPM input command message blocks. When the authorization block is complete, build the complete TPM_Sign command by

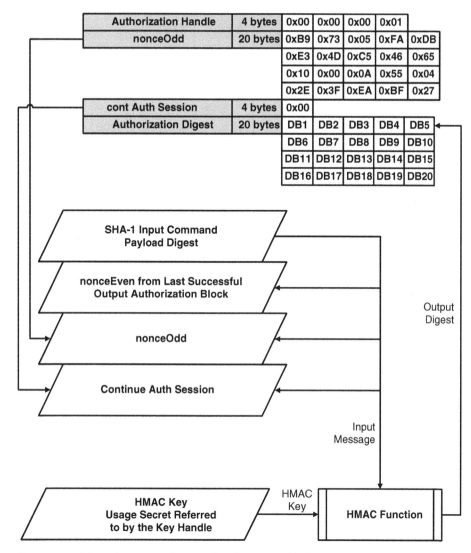

Figure 9.21 TPM_Sign input digest and authorization block

concatenating the TPM_Sign input command block with the authorization block. Figure 9.22 outlines the full TPM_Sign command.

Let's look at one interesting piece of information regarding the authorization block parameter and continue authorization session. Notice that the TPM_LoadKey has this parameter set to one (1). This says to the TPM that the TPM_OIAP authorization session is to remain open and the session will remain open unless the command fails.

Next notice the continue authorization session parameter relative to the TPM_Sign authorization block; it is defined as a zero (0). This tells the TPM to

Authorization Tag	2 bytes	0x00	0xC2			
Parameter Size	4 bytes	0x00	0x00	0x00	0x53	
Ordinal	4 bytes	0x00	0x00	0x00	0x3C	
Key Handle	4 bytes	0x00	0x00	0x00	0x01	
Size of Message	4 bytes	0x00	0x00	0x00	0x14	
Message to Sign	20 bytes	0xB3	0xD5	0xCB	0x12	0x73
		0x8B	0xB6	0xF9	0x21	0xA3
		0xDA	0x42	0xE0	0x18	0xD1
		0x43	0xFA	0x29	0x7C	0xA6
Authorization Handle	4 bytes	0x00	0x00	0x00	0x01	
nonceOdd	20 bytes	0xB3	0x73	0x05	0xFA	0xDB
		0xE3	0x4D	0xC5	0x46	0x65
		0x10	0x00	0x0A	0x55	0x04
		0x2E	0x3F	0xEA	0xBF	0x27
Continue Auth Session	1 bytes	0x00				
Authorization Digest	20 bytes	DB1	DB2	DB3	DB4	DB5
		DB6	DB7	DB8	DB9	DB10
		DB11	DB12	DB13	DB14	DB15
		DB16	DB17	DB18	DB19	DB20

Figure 9.22 The TPM_Sign input message block with authorization

close the OIAP authorization session after successful execution of the Sign command. This is an example of an implicit authorization session termination as opposed to the explicit command execution concerning the TPM_Terminate_ Handle. Also note that the session will be terminated after the TPM_Sign output message and authorization block has been transmitted back to the TSS. I know, an obvious statement, but it had to be written.

So the command executed as expected and the TSS has received the corresponding TPM_Sign output message – time to validate the authorization digest without going through the complete detail. If you need to, refer to the discussion concerning the TPM_LoadKey output authorization in the previous example.

Figure 9.23 outlines the authorization digest calculation concerning the TPM_Sign output message. As before, calculate the authorization digest and compare the calculated value to the authorization digest contained in the authorization block. On successful comparison, the signature can now be stored in the TSS or used for whatever reason for which this signature was generated.

One point of interest concerns the TPM_Sign output message length: it is variable depending on the size of the key material used to sign the digest.

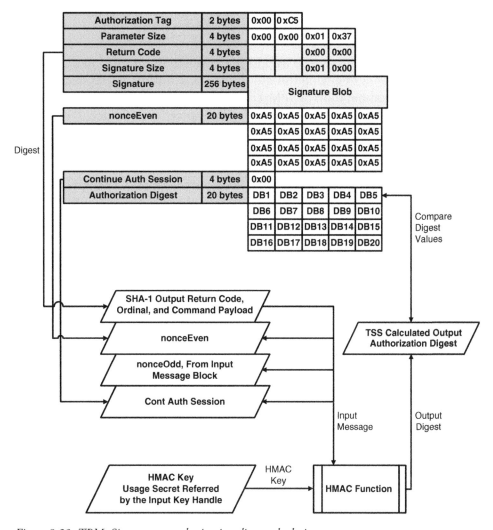

Figure 9.23 TPM_Sign output authorization digest calculation

Most TPM's support various public key sizes including 512, 1024, and 2048 bits. So if the signing key is 512 bits, the signature will be 64 bytes; 1024 bits, the signature will be 128 bytes; and 2048 bits, the signature will be 256 bytes. This is important if you are designing a "home-brewed" TSS or Embedded TSS: watch out for varying input and output message lengths that have a dependency concerning keying material sizing. Another little piece on information while we're on this subject: this is not the only dependency that will cause varying message lengths. Just keep that information in the back of your mind as we progress through the rest of the chapters.

Example 2: Load Key and Sign Using an OSAP Session

Let's make this as easy as possible; instead of introducing another set of commands depicting the use of the OSAP authorization session, we'll use the same combination of TPM_LoadKey and TPM_Sign commands and replace the OIAP with an OSAP. To put a finer point on this idea: the OSAP will only apply to the TPM_Sign command. Any guesses why? Come on think about it, OSAP. OK, the TPM_LoadKey, in the example that we designed, uses the SRK as the parent key and the TPM_Sign command uses the loaded key – the product of the load key command to sign the SHA-1 digest. The quick answer is: we are working with two separate objects here, the SRK and the Signing Key, so I am picking the Signing Key to use an OSAP session.

First, perform the same load key operation, as defined in Example 1, that will load the Signing Key into the TPM and return a handle to the keying material held within the TPM. A point of concern: before doing this make sure that there are no current authorization sessions defined within the TPM. The quickest method of doing this is to execute two TPM_Terminate_Handle commands, and if in compliance mode, this would be the handle values 0x00000000 and 0x00000001. If, by chance, you're in a noncompliance state, then the best solution would be to power cycle the TPM and reinitialize the TPM by executing the TPM startup command sequence defined in previous chapters. The reason for this drastic sequence, in the case of a noncompliance TPM operation, is the fact that the noncompliance state uses random handles; if you didn't save the handles, you can't possibly terminate the session. Sorry, you loose.

Once the key is loaded and the Key Handle is known, the OSAP session can be created in anticipation of the pending TPM_Sign command execution. With that said, let's get going on the input command message block needed to create the OSAP authorization session within the TPM. Referring to the previous paragraph, create the Input OSAP message block defined there. For example,

Authorization Tag	2 bytes	0x00	0xC1			
Parameter Size	4 bytes	0x00	0x00	0x00	0x24	
Command Ordinal	4 bytes	0x00	0x00	0x00	0x0B	
Entity Type	2 bytes	0x00	0x05			
Entity Value	4 bytes	0x00	0x00	0x00	0x02	
nonceOSAPOdd	4 bytes	0xB9	0x73	0x05	0xFA	0xDB
		0xE3	0x4D	0xC5	0x46	0x65
		0x10	0x00	0x0A	0x55	0x04
		0x2E	0x3F	0xEA	0xBF	0x27

Figure 9.24 TPM_OSAP input message block

Figure 9.24 outlines a specific input message block that could be sent to the TPM within the compliance state.

First the usual tag, parameter size, and ordinal values are populated. Then the new parameters in which we have been exposed to the Entity Type, the Entity Value, and the nonceOddOSAP values. Start with the Entity Type, in this case we are using a Key Handle, so this parameter takes on the TCG defined value of TCG_ET_KEYHANDLE. Next is the value of the Key Handle in question and this will be defined as the Key Handle that was passed back via the output message of the TPM_LoadKey command. Finally, just like the nonceOdd parameter, generate a random, in this case, a compliance mode (not so random) with a 20-byte value, and assign this value to be the nonceOddOSAP. *Note*: do not forget this value; if you are writing test bench applications, store this value because you will need it later in the game. In the case of compliance mode this particular point is not so interesting; but when you migrate to noncompliance vectors, you will feel the pain if you do not take heed and save the nonceOddOSAP.

When this command is executed you will get back the output message associated with the TPM_OSAP command as previously described. Figure 9.25 defines a specific output message block that could be associated with a successful OSAP command execution. This command output message block is very similar to the OIAP command message block, the usual culprits: tag, parameter size, return code, authorization handle, and nonceEven; however, there is another 20-byte parameter, the nonceEvenOSAP. *Important point*: the nonceEven is used to authorize command-level operations and the nonceEvenOSAP is used to generate the Shared Secret that will be the HMAC key when generating the authorization digest associated with the command and tied to the entity, in this case a signing key.

Authorization Tag	2 bytes	0x00	0xC4			
Parameter Size	4 bytes	0x00	0x00	0x00	0x36	
Return Code	4 bytes	0x00	0x00	0x00	0x00	
Authorization Handle	4 bytes	0x00	0x00	0x00	0x01	
nonceEven	20 bytes	0xA5	0xA5	0xA5	0xA5	0xA5
		0xA5	0xA5	0xA5	0xA5	0xA5
		0xA5	0xA5	0xA5	0xA5	0xA5
		0xA5	0xA5	0xA5	0xA5	0xA5
nonceEvenOSAP	20 bytes	0xA5	0xA5	0xA5	0xA5	0xA5
		0xA5	0xA5	0xA5	0xA5	0xA5
		0xA5	0xA5	0xA5	0xA5	0xA5
		0xA5	0xA5	0xA5	0xA5	0xA5

Figure 9.25 TPM_OSAP output message block

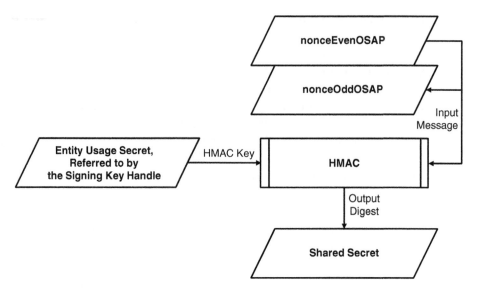

Figure 9.26 Generation of Shared Secret

The first order of business is to generate the Shared Secret, which will become the HMAC key for both the input and output message authorization digest value generation. Figure 9.26 depicts the Shared Secret generation specific to this example.

Notice that the HMAC key is the usage authorization value associated with the signing key that we loaded into the TPM. This is an important concept because the OSAP is tied to this keying material specifically; and if we tried to use this OSAP session in the authorization of another signature operation that leveraged a different signing key, the command would fail. The reason for this command failure concerns the usage authorization they are different in the case of two separate signature keys and therefore the Shared Secret would differ because of the dependency on the HMAC key. Therefore the specific dependency is an implicit one whose root concerns the signature key's Usage Secret and is enforced by the Shared Secret when calculating authorization digest values used in the command authorization protocol.

After the Shared Secret is calculated, it replaces the Usage Secret that was used with regard to the TPM_Sign command HMAC key that produced the authorization digest as defined in Example 1. Now we can design an input command block with an attached authorization block specifically tailored to the TPM_Sign command.

Figure 9.27 defines the authorization block with the authorization digest calculation, which will be concatenated with the specific command parameters.

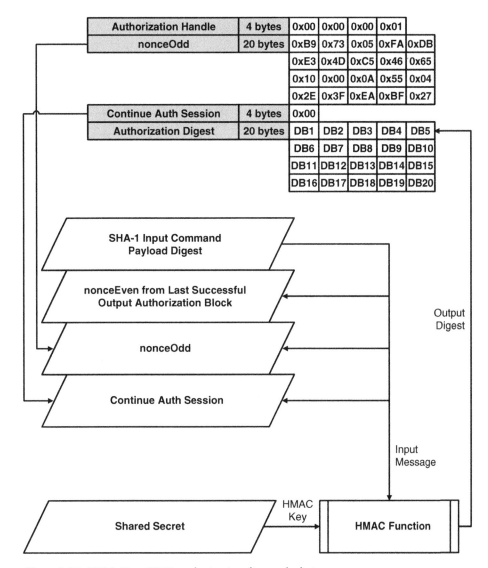

Authorization Handle	4 bytes	0x00	0x00	0x00	0x01
nonceOdd	20 bytes	0xB9	0x73	0x05	0xFA	0xDB
		0xE3	0x4D	0xC5	0x46	0x65
		0x10	0x00	0x0A	0x55	0x04
		0x2E	0x3F	0xEA	0xBF	0x27
Continue Auth Session	4 bytes	0x00
Authorization Digest	20 bytes	DB1	DB2	DB3	DB4	DB5
		DB6	DB7	DB8	DB9	DB10
		DB11	DB12	DB13	DB14	DB15
		DB16	DB17	DB18	DB19	DB20

Figure 9.27 TPM_Sign OSAP authorization digest calculation

Notice the dependency concerning the authorization digest to the Shared Secret instead of the Usage Secret, as defined in Example 1.

Figure 9.28 defines the entire TPM_Sign input command with the authorization block attached; this is a complete input message command that can be sent to the TPM for execution and is associated with an OSAP authorization protocol. An item of interest would be to replace the OSAP authorization handle with the OIAP authorization handle that was used to load the signing key into the TPM. The Signature command would fail, in this

Authorization Tag	2 bytes	0x00	0xC5			
Parameter Size	4 bytes	0x00	0x00	0x00	0x53	
Ordinal	4 bytes	0x00	0x00	0x00	0x3C	
Key Handle	4 bytes	0x00	0x00	0x01	0x00	
Size of Message	4 bytes	0x00	0x00	0x00	0x14	
Message to Sign	4 bytes	0xB3	0xD5	0xCB	0x12	0x73
		0x8B	0xB6	0xF9	0x21	0xA3
		0xDA	0x42	0xE0	0x18	0xD1
		0x43	0xFA	0x29	0x7C	0xA6
AuthHandleOSAP	4 bytes	0x00	0x00	0x00	0x01	
nonceOdd	20 bytes	0xB3	0x73	0x05	0xFA	0xDB
		0xE3	0x4D	0xC5	0x46	0x65
		0x10	0x00	0x0A	0x55	0x04
		0x2E	0x3F	0xEA	0xBF	0x27
Continue Auth Session	4 bytes	0x00				
Authorization Digest	20 bytes	DB1	DB2	DB3	DB4	DB5
		DB6	DB7	DB8	DB9	DB10
		DB11	DB12	DB13	DB14	DB15
		DB16	DB17	DB18	DB19	DB20

Figure 9.28 OSAP-based TPM_Sign input message block

case, because the OIAP authorization protocol would leverage the Usage Secret and not the Shared Secret that the OSAP protocol defines. Remember, the first authorization session in this example is defined as an OIAP and the second authorization session is defined as an OSAP. Each of the sessions defines very specific protocols that determine how the authorization digest will be calculated.

Moving on, Figure 9.29 defines the TPM_Sign output execution block that will be returned in lieu of a successful command execution. Note that the same message parameters apply to this output message block as in the TPM_Sign output message block in Example 1. The authorization block has been added to this figure to expedite the discussion concerning authorization digest verification.

Speaking of which, the procedure for verifying the output authorization digest is exactly the same as defined in Example 1, with the only exception being the HMAC key. In Example 1, the HMAC key was the Usage Secret associated with the signing key; in this example, the HMAC key is the Shared

Authorization Tag	2 bytes	0x00	0xC5			
Parameter Size	4 bytes	0x00	0x00	0x01	0x37	
Return Code	4 bytes	0x00	0x00	0x00	0x00	
Signature Size	4 bytes	0x00	0x00	0x01	0x00	
Signature	256 bytes					

Signature Blob

nonceEven	20 bytes	0xA5	0xA5	0xA5	0xA5	0xA5
		0xA5	0xA5	0xA5	0xA5	0xA5
		0xA5	0xA5	0xA5	0xA5	0xA5
		0xA5	0xA5	0xA5	0xA5	0xA5
Continue Auth Session	4 bytes	0x00				
Authorization Digest	20 bytes	DB1	DB2	DB3	DB4	DB5
		DB6	DB7	DB8	DB9	DB10
		DB11	DB12	DB13	DB14	DB15
		DB16	DB17	DB18	DB19	DB20

Figure 9.29 OSAP-based TPM_Sign output message block

Secret associated with the OSAP session that has an implicit dependency on the Usage Secret relative to the signing key. Figure 9.30 defines the authorization validation procedure.

In summary, command authorization is accomplished through two authorization protocols – the OIAP and the OSAP. Some commands require the use of the OSAP to authorize their execution and other commands can use the OSAP instead of the OIAP, depending on the context relative to command execution. The important points to remember are as follows:

- The OIAP uses the Usage Secret directly with regard to the HMAC key.
- The OSAP leverages the nonceOddOSAP and nonceEvenOSAP.
- The OSAP generates the Shared Secret and uses this as the HMAC key.
- The OSAP is implicitly tied to the entity defined within the generated session.
- The OIAP is independent of any entity since it uses the Usage Secret explicitly.

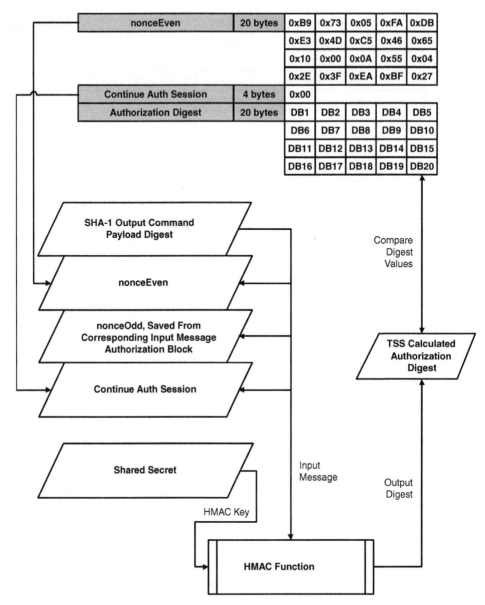

Figure 9.30 Validating the TPM_Sign OSAP authorization digest

Termination of the authorization session can be accomplished implicitly via the continue authorization session contained within the input message block or explicitly by executing the TPM_Terminate_Handle command.

10

Command Authorization, Atypical

This chapter defines the exceptions concerning the rule(s) in place regarding command authorization. The two most important concepts about authorization protocol rule exceptions involve the subject of deferred authorization and the non-authorized execution of a normally authorized command. The first to be examined is the deferred authorization protocol as applied to the Trusted Computing Group (TCG) command TPM_TakeOwnership. Note that the specific command details will be glossed over for the specific purpose of understanding the deferred authorization concept. In Chapter 25, the TPM_TakeOwnership command is detailed in its full glory, but as always, you can skip ahead and familiarize yourself with this command's specifics.

10.1 Exception Case, the Deferred Authorization Protocol

What exactly is a deferred authorization and way would any command need the functionality that this authorization type portrays? To better understand this, let's look at the Trusted Platform Module (TPM) state after the compliance vectors have been cleared and an Endorsement Key (EK) pair has been created. For those of you scratching your heads wondering how this state was acquired, be patient; we are only discussing why a deferred authorization is supported. With the TPM having only the EK pair – the RSA keying material that attests to the authenticity of the physical TPM device – the TPM has no Owner established and no Storage Root Key (SRK) to leverage any Owner cryptographic functionality.

This poses a slight problem concerning the previous chapter's examples, which leveraged the SRK to load in alternate keying material. Just for exploratory purposes, the SRK is the base key in the key management tree that allows other keys to be loaded into the TPM or generated internally in the TPM. To complete an explanation regarding the deferred authorization, Figure 10.1 defines a basic, and entirely possible, key management tree that could exist within the boundaries of the TPM physical device.

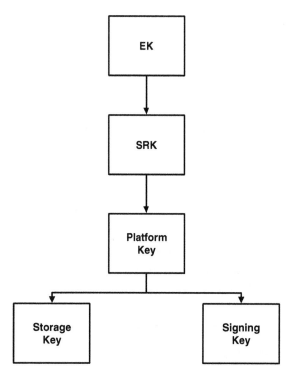

Figure 10.1 Possible key management tree in the TPM

Notice that the EK is depicted as the top key in the key management structure. This can be a very deceptive concept regarding the cryptographic functionality that can be leveraged by the TPM Command Suite with regard to this key pair. In short, the EK provides the root-of-trust and the ability to securely execute the TPM_TakeOwnership command. Therefore, if you want to prove that the TPM produced by, for example, the Atmel Corporation, the EK public key can be used to provide certification, indeed this TPM is a valid Atmel TPM. Another function the EK provides is the ability to secure the private information that will exist and be protected by the TPM with regard to the Owner Secret and the SRK Usage Secret. One problem exists: there is no parent key loaded into the TPM that will facilitate the TPM_LoadKey concerning the SRK, and this is where the deferred authorization comes into play. For more detail concerning the subject of key hierarchy refer to Chapter 15 about key management.

Let's suppose you could send a command to the TPM without performing an authorization check, and after a portion of that command has been executed, you were given the opportunity to validate its authorization digest. This little blurb just defined what it is to perform a deferred authorization on an input command and the example that will portray this concept is the TPM_TakeOwnership command. The TakeOwnership command is executed during a TPM state that has only one available cryptographic key pair available, the EK Pair. This means that any data to be protected during the input message transmission relative to the TakeOwnership command must be encrypted by the EK public key. The two pieces of data that must be secure during the transmission for this command input block concern the Owner Secret and the SRK Usage Secret. Also note that to successfully authorize the TPM_TakeOwnership command, the TPM must have knowledge of the Owner Secret. The very technical term regarding this problem is referred to as "the chicken and the egg" and which of them came first.

Therefore, the TCG Software Stack (TSS) must encrypt the Owner Secret and the SRK Usage Secret with the EK public key, which will always produce a 256-byte value for each. This is due to the fact that the EK is a 2048-bit RSA key pair, and makes you wonder why the TCG uses two parameters to define the size of the encrypted secrets; one may never know. After the secrets are transformed into 256-byte blobs, the rest of the TakeOwnership command can be completed. The resulting input message block is defined in Figure 10.2.

The next task concerning the TSS is to create an authorization block that will successfully authorize the TPM_TakeOwnership input command message. First, this example depends on the successful execution of the TPM_OIAP command since we must have an internal authorization cache and authorization handle to that cache in order to execute an authorized command. The previous example (see Chapter 9), which describes the TPM_OIAP command execution, should be a sufficient explanation of how to create these authorization dependencies.

After the authorization cache is created, the method for calculating the authorization digest is identical to previous examples with one exceptions the Hash-based Message Authentication Code (HMAC) key needs to be the

Authorization Tag	2 bytes	0x00	0xC2		
Parameter Size	4 bytes	0x00	0x00	0x02	0x70
Ordinal	4 bytes	0x00	0x00	0x00	0x0D
Protocol ID	2 bytes	0x00	0x05		
Enc Owner Auth Size	4 bytes	0x00	0x00	0x01	0x00
Enc Owner Auth	256 bytes	Encrpted Blob Protecting the Owner Authorization			
Enc SRK Auth Size	4 bytes	0x00	0x00	0x01	0x00
Enc SRK Auth	256 bytes	Encrpted Blob Protecting the SRK Authorization			
TCPA_KEY Structure	4 bytes	0x01	0x01	0x00	0x06
		0x00	0x11		
		0x00			
		0x00	0x00	0x00	0x01
		0x00	0x00	0x00	0x01
		0x00	0x03		
		0x00	0x01		
		0x00	0x00	0x00	0x0C
		0x00	0x00	0x80	0x00
		0x00	0x00	0x00	0x02
		0x00	0x00	0x00	0x00
		0x00	0x00	0x00	0x00
		0x00	0x00	0x00	0x00
		0x00	0x00	0x00	0x00
		0x00	0x00	0x00	0x01

Figure 10.2 TPM_TakeOwnership input message block

Owner Secret previously encrypted by the EK public key. Figure 10.3 defines the TPM_TakeOwnership authorization digest calculation and puts this digest within the context of the authorization message block, which will be attached to the command message block to form the complete input message.

That's all folks. The next step is to transmit the TPM_TakeOwnership owner-authorized command to the TPM for execution. The TPM will now chew on the data, and here is where things get interesting. Remember the

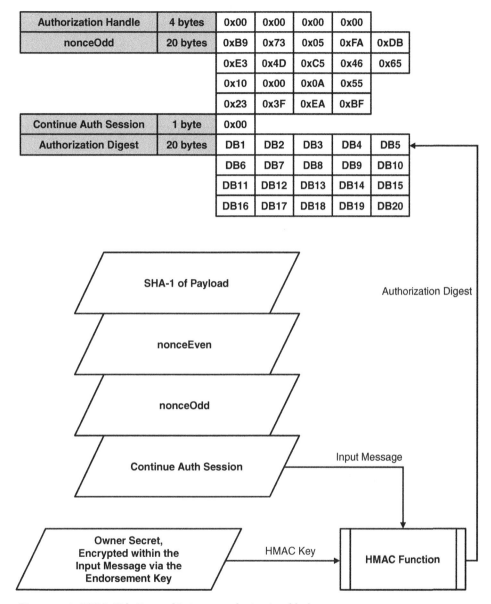

Authorization Handle	4 bytes	0x00	0x00	0x00	0x00	
nonceOdd	20 bytes	0xB9	0x73	0x05	0xFA	0xDB
		0xE3	0x4D	0xC5	0x46	0x65
		0x10	0x00	0x0A	0x55	
		0x23	0x3F	0xEA	0xBF	
Continue Auth Session	1 byte	0x00				
Authorization Digest	20 bytes	DB1	DB2	DB3	DB4	DB5
		DB6	DB7	DB8	DB9	DB10
		DB11	DB12	DB13	DB14	DB15
		DB16	DB17	DB18	DB19	DB20

Figure 10.3 TPM_TakeOwnership input authorization block

TPM's authorization engine does not have all the needed data to calculate the authorization digest and the Usage Secret is not known. Hence, the TPM must perform a deferred authorization protocol, that is, defer the command authorization until the entity's Usage Key is available – in the case, until the Owner Secret has been decrypted. The steps taken during the TakeOwnership command execution are defined graphically in Figure 10.4.

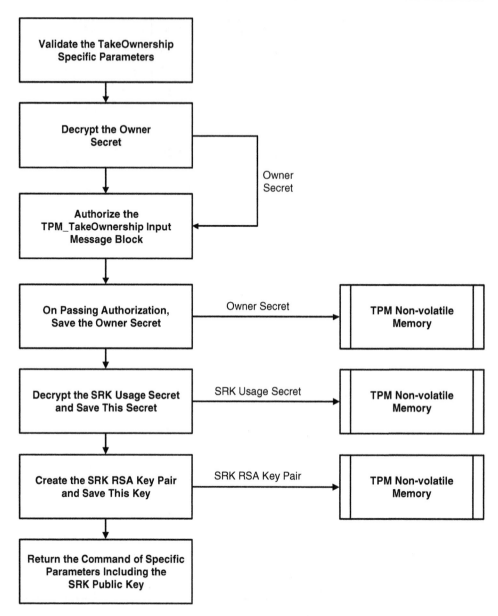

Figure 10.4 Functional steps taken during TPM ownership creation

The steps defined here are supplied at the 30,000-foot view and are only defined down to a level of detail that will help readers understand the deferred authorization protocol. The first step is to decrypt the Owner Secret in order to authorize the TPM_TakeOwnership command. The TPM decrypts the 20-byte secret using the EK private key; remember, the secret was encrypted using the EK public key. Once this is done, the TPM can compute the TPM_TakeOwnership command's authorization digest and repeat the algorithm that the TSS performed in calculating this value. The TPM calculates and validates the authorization digest and only then will it store the Owner Secret within its non-volatile memory for future owner-authorized commands. The same procedure is taken concerning the SRK Usage Secret; the only difference is that there is no validation required by TPM. The TPM simply decrypts the SRK Usage Secret and stores this value in its non-volatile memory for future SRK authorization use. Finally, the TPM creates the SRK keying material, as defined by the key data passed in via the TPM_TakeOwnership input command parameters, and stores this key within the TPM for future cryptographic use.

Figure 10.5 defines the output message block that is returned to the TSS for validation and consumption purposes. This figure shows the output authorization calculation performed internally in the TPM and externally by the TSS validating the command output transaction. There should be no surprises here regarding the calculation and validation of the authorization digest that is tied to this command's output message block.

The point of this discussion is to make readers aware of the concept concerning deferred authorization and to put a real-world viewpoint on the subject matter. Understanding the deferred authorization concept early on, hopefully, will free everyone from this burden when examining the details and designing test applications, which leverage commands making use of the particular authorization protocol.

10.2 Exception Case, Non-authorized Command Execution of Normally Authorized Commands

Wow, that was a mouthful. Quite a few TPM commands are of the authorized flavor as defined within their natural state of execution by Version 1.1b of the TCG Main Specification. This section describes a situation that exhibits the exception to the rule using our old buddy, the TPM_LoadKey command. Remember that the TPM_LoadKey command is defined as a single authorized command and all previous examples were leveraged as such. In this example, we are going to "override" the normal authorization requirement with regard to TPM_LoadKey and execute this command as non-authorized. That's not to

Authorization Tag	2 bytes	0x00	0xC4		
Parameter Size	4 bytes	0x00	0x00	0x01	0x62
Return Code	4 bytes	0x00	0x00	0x00	0x00
TCPA_KEY Structure	4 bytes	0x01	0x01	0x00	0x06
		0x00	0x11		
		0x00			
		0x00	0x00	0x00	0x01
		0x00	0x00	0x00	0x01
		0x00	0x03		
		0x00	0x01		
		0x00	0x00	0x00	0x0C
		0x00	0x00	0x80	0x00
		0x00	0x00	0x00	0x02
		0x00	0x00	0x00	0x00
		0x00	0x00	0x00	0x00
		0x00	0x00	0x01	0x00
		SRK Public Key			
		0x00	0x00	0x00	0x00
nonceEven	20 bytes	0xA5	0xA5	0xA5	0xA5	0xA5
		0xA5	0xA5	0xA5	0xA5	0xA5
		0xA5	0xA5	0xA5	0xA5	0xA5
		0xA5	0xA5	0xA5	0xA5	0xA5
Continue Auth Session	1 byte	0x00			
Authorization Digest	20 bytes	DB1	DB2	DB3	DB4	DB5
		DB6	DB7	DB8	DB9	DB10
		DB11	DB12	DB13	DB14	DB15
		DB16	DB17	DB18	DB19	DB20

Figure 10.5 TPM_TakeOwnership output message block

say that we can ignore the TPM authorization requirements; it just means that we are going to "set up" the TPM state to allow us to execute a non-authorized TPM_LoadKey command. As you will see, there are very specific steps that are required, which are very much done within a secure environment.

The first step is to create a parent key – the key used to encrypt the private data within the TPM_LoadKey input command and has the no authorization property invoked. For a quick peek ahead, any key can be loaded within the TPM with the specific instruction that its cryptographic usage be non-authorized. This property is set within the key parameters, which are passed on to the TPM via the TPM_LoadKey command input message. To facilitate this, a second storage key will be loaded below the SRK – the SRK will be its parent and this key will be defined as a non-authorized key. Figure 10.6 shows the key management tree of the newly loaded key relative to the EK and SRK.

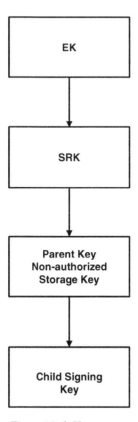

Figure 10.6 Key management of the non-authorized storage key

With this goal in mind, we will now define the TPM command that will load this key into the TPM and allow us to leverage it in performing a non-authorized TPM_LoadKey command. First, this example is dependent on the SRK residing within the TPM and that the owner has been established. The TPM_LoadKey will be used to load the non-authorized keying material with the intent of executing a second TPM_LoadKey using the non-authorized key as a parent. Figure 10.7 outlines the TPM_LoadKey input command message that will perform the loading of a non-authorized key.

Authorization Tag	2 bytes	0x00	0xC2		
Parameter Size	4 bytes	0x00	0x00	0x02	0x6a
Ordinal	4 bytes	0x00	0x00	0x00	0x20
Key Handle	4 bytes	0x40	0x00	0x00	0x00
TCG Key Structure	555 bytes	0x01	0x01	0x00	0x00
		0x00	0x10		
		0x00	0x00	0x00	0x02
		0x00			
		0x00	0x00	0x00	0x01
		0x00	0x01		
		0x00	0x03		
		0x00	0x00	0x00	0x0c
		0x00	0x00	0x80	0x00
		0x00	0x00	0x00	0x02
		0x00	0x00	0x80	0x00
		0x00	0x00	0x80	0x00
		0x00	0x00	0x01	0x00
		Encrypted Blob Reference Key Handle			
		0x00	0x00	0x01	0x00
		Encrypted Blob Reference Key Handle			
Authorization Handle	4 bytes	0x00	0x00	0x00	0x00
nonceOdd	20 bytes	0xB9	0x73	0x05	0xFA	0xDB
		0xE3	0x4D	0xC5	0x46	0x65
		0x10	0x00	0x0A	0x55	0x04
		0x2E	0x3F	0xEA	0xBF	0x27
Continue Auth Session	4 bytes	0x00			
Authorization Digest	20 bytes	DB1	DB2	DB3	DB4	DB5
		DB6	DB7	DB8	DB9	DB10
		DB11	DB12	DB13	DB14	DB15
		DB16	DB17	DB18	DB19	DB20

Figure 10.7 TPM_LoadKey input message loading a non-authorized key

Looking at the input message block it is not very apparent as to how this TPM_LoadKey differs from the previous examples. The short answer is the TCG_AUTH_DATA_USAGE parameter that is a sub-parameter within the larger TCG_KEY structure embedded within the TPM_LoadKey input command. The specific definition regarding the TCG_AUTH_DATA_USAGE parameter is shown in Figure 10.8.

TPM_AUTH_NEVER	0x00
TPM_AUTH_ALWAYS	0x01

Figure 10.8 TCG authorization data usage types

In Figure 10.7, a single byte is highlighted, which identifies the TCG_AUTH_DATA_USAGE parameter. Notice that its value is defined as 0x00, indicating that the key being load requires no Usage Secret or authorization to gain access with regard to the key's cryptographic functionality. That is, very simply, 1 byte is all that is needed to modify the authorization property of a key that is about to be loaded into the TPM. After successful execution, the TPM has a non-authorized key available for use that will allow a non-authorized TPM_LoadKey execution.

Now's the time we've been all waiting for – the non-authorized TPM_Load Key example. The non-authorized input command will look much different in comparison to its authorized cousin. Figure 10.9 defines the non-authorized TPM_LoadKey input command message.

The first item of interest concerns the authorization tag; its value is 0x0001 or no authorization required. In addition, there is no authorization block attached to the TPM_LoadKey input message as opposed to previous examples. Executing the non-authorized TPM_LoadKey is no different from its authorized version. The output message produced from the execution of the non-authorized TPM_LoadKey command varies as well compared to its authorized version. Figure 10.10 defines the output message specifics.

Notice something. That's right, there is no authorization block attached to the output message and the authorization tag value is 0x00C4, indicating a non-authorized TPM output message. No surprises here. The big picture is that most – though there are commands that can't have their authorization levels changed – commands allow the TPM to be put in a state that overrides their specified authorization level. This is due to the fact that the authorization

Authorization Tag	2 bytes	0x00	0xC1		
Parameter Size	4 bytes	0x00	0x00	0x02	0x6a
Ordinal	4 bytes	0x00	0x00	0x00	0x20
Key Handle	4 bytes	0x40	0x00	0x00	0x01
TCG Key Structure	555 bytes	0x01	0x01	0x00	0x00
		0x00	0x10		
		0x00	0x00	0x00	0x02
		0x01			
		0x00	0x00	0x00	0x01
		0x00	0x01		
		0x00	0x03		
		0x00	0x00	0x00	0x0c
		0x00	0x00	0x80	0x00
		0x00	0x00	0x00	0x02
		0x00	0x00	0x80	0x00
		0x00	0x00	0x80	0x00
		0x00	0x00	0x01	0x00
		Encrypted Blob Reference Key Handle			
		0x00	0x00	0x01	0x00
		Encrypted Blob Reference Key Handle			

Figure 10.9 Non-authorized TPM_LoadKey input message

Authorization Tag	2 bytes	0x00	0xC4		
Parameter Size	4 bytes	0x00	0x00	0x00	0x0d
Return Code	4 bytes	0x00	0x00	0x00	0x00
Key Handle	4 bytes	0x00	0x00	0x00	0x04

Figure 10.10 Non-authorized TPM_LoadKey output message

level is determined by the entity that causes the authorization level in the first place. For example, the TPM_LoadKey command is authorized due to the fact that this command leverages a RSA key to protect the loaded key's private data. The RSA key usually requires a known Usage Secret used to authorized its use; hence, the reason the TPM_LoadKey is authorized to begin with. If the RSA key is declared or loaded with the TCG_AUTH_DATA_USAGE parameter defined as non-authorized, there is no reason the TPM_LoadKey command needs authorization by the TPM. If you like, take a look at some other commands and decide for yourself why the command is authorized and, if so, can this command be modified to execute as a non-authorized.

10.3 Exception Case, the EncAuth

EncAuth, what the heck is an **EncAuth**? Good question. I have been sort of wondering that myself. I really don't know why some of the principles defined by the TCG have such obscure names, but my goal is not to question, it's to explain. First of all, let's get this idea under control – the idea of protecting a secret without making use of an RSA key. We have seen examples were private data has been encrypted by some RSA key to protect information being transmitted to the TPM – sure we have. Well, there is another method invoked in regard to protecting information, the infamous **EncAuth**. Let's get something straight right away; the **EncAuth** is nothing more than an eXclusive OR (XOR) function, that's it. Instead of encrypting, let's say a Usage Secret using an RSA public key, we XOR the Usage Secret with a 20-byte pad. That's it, the **EncAuth** in all of its glory or obfuscation. You make the call.

There are two specific goals concerning the **EncAuth**: create the 20-byte pad and perform the XOR function. Specific examples will not be leveraged to explain the **EncAuth** functionality, but if you look at a command such as TPM_ChangeAuth, you will see the **EncAuth** in action. The priority here is to understand and use the **EncAuth** protocol so that when you come up against it, you will know its purpose.

Moving on. Let's define the procedure regarding the creation of the 20-byte pad used by the XOR function when decrypting and encrypting data. I use the terms encrypt and decrypt very loosely within this context. Figure 10.11 defines the data and operations that must be performed in the generation of the 20-byte pad. As you can see, this is a simple function. The simple Secure Hash Algorithm (SHA-1) hashes the concatenation of the Shared Secret. **EncAuth** requires an Object Specific Authorization Protocol (OSAP) session with the LastnonceEven.

What do I mean by LastnonceEven? Remember that the nonceEven and nonceOdd are "rolling nonces" and at the generation of the output message, the TPM creates a new nonceEven. The LastnonceEven is the nonceEven prior to generating at new nonceEven, or simply put, the nonceEven associated with the command's input message block.

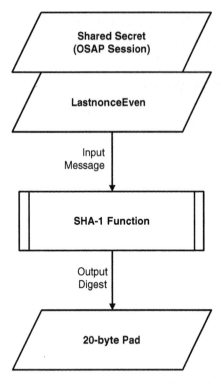

Figure 10.11 Generating the 20-byte pad, EncAuth

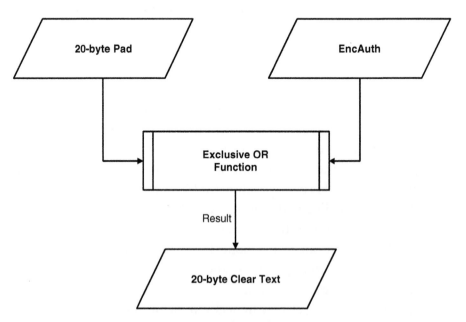

Figure 10.12 XOR the EncAuth with the 20-byte pad

After the 20-byte pad has been created, the last step is to perform the XOR function. Figure 10.12 defines the functional description concerning this task. This is a very simple detail. When the XOR function has been executed, the result is the clear text – in this case, the 20-byte usage authorization.

One thing that is worth mentioning: the **EncAuth** is somewhat similar to a symmetric key. That is, the same key is used to encrypt as well as decrypt data, as opposed the RSA – an asymmetric cryptosystem – that uses a public and a private key pair.

In summary, the EncAuth allows 20-byte secrets, or data, that needs to be protected to be encrypted using a method other than RSA. One reason for this involves the size of the resulting cipher text when comparing RSA to **EncAuth**. For example, the **EncAuth** takes a 20-byte clear text and produces a 20-byte cipher text. RSA, on the other hand, takes a 20-byte clear text and produces 64, 128, or 256 bytes of cipher text depending on the key size – 512, 1024, or 2048 bits.

11

The Initialization and Low-Level Command Suite

There are two basic forms of Trusted Platform Module (TPM) initialization: one used during application development based on compliance vectors and one used to prepare the TPM for field deployment. Given that different variations can be employed regarding TPM development, this chapter focuses on TPM initialization using the compliance vectors – the vectors present in the TPM at the time of delivery. One reason for this choice is because of the example chapters later in this book.

Cryptography involves systems of high entropy and defining predictable results would, to say the least, undermine the benefit that any cryptographic system provides. On the other hand, providing predictable results is a huge benefit when developing applications leveraging the functionality that the TPM provides. Therefore, this chapter is devoted to initializing the TPM so as to utilize the compliance vectors, which will give predictable results to TPM command execution. Chapter 23, TPM System Deployment Initialization, describes the TPM initialization procedure with regard to embedded systems about to be deployed into the field.

11.1 Determining TPM Compliance State

One problem that is faced immediately when developing applications that have a dependency concerning the TPM with compliance vectors is the determination that these vectors are indeed available. Looking at the volatile and non-volatile TPM flags, none of these state variables defines the presence of compliance vectors. Indeed, looking over Version 1.1b of the Trusted Computing Group (TCG) Main Specification, you would be hardpressed to find any description concerning the presence of compliance vectors in the TPM. So there must be a simple method for ascertaining the presence of these vectors and there is. In fact there are many ways to determine the answer to the preceding problem, but the way that I have done this is to execute a TPM Object Independent Authorization Protocol (OIAP) command and look at the nonceEven produced within the output message block.

Figure 11.1 defines the actual command byte sequence that will be used during the compliance vector test. This is your simple TPM_OIAP command, but the interesting data is contained within the command output message block.

Authorization Tag	0x00	0xC1		
Parameter Size	0x00	0x00	0x00	0x0A
Ordinal	0x00	0x00	0x00	0x0A

Figure 11.1 Determining compliance vector inclusion – TPM_OIAP input message

This example assumes that your TPM device has compliance vectors resident and that the TPM_OIAP command output message block will produce the response defined in Figure 11.2.

Authorization Tag	0x00	0xC4			
Parameter Size	0x00	0x00	0x00	0x22	
Return Code	0x00	0x00	0x00	0x00	
Authorization Handle	0x00	0x00	0x00	0x00	
nonceEven	0xA5	0xA5	0xA5	0xA5	0xA5
	0xA5	0xA5	0xA5	0xA5	0xA5
	0xA5	0xA5	0xA5	0xA5	0xA5
	0xA5	0xA5	0xA5	0xA5	0xA5

Figure 11.2 Determining compliance vector inclusion – TPM_OIAP output message

Assuming that the command execution was successful – there were no resource errors – the nonceEven produced by the TPM is a constant value; a 20-byte nonce of 0xA5's. This is the compliance definition for the nonceEven produced by the TPM and when in the compliance state, the TPM will always produce this pattern for the TPM nonceEven value. If, on the other hand, the TPM nonceEven is a random value of 20-bytes, you can be assured that the compliance vectors are not present and you will not be able to leverage predictable results regarding TPM command execution. In addition, once the compliance vectors have been "cleared", by executing one of the variations concerning the TPM force clear commands, you will never be able to restore these values to that TPM.

The moral of this story is to be confident that you will not have the need for the compliance vectors prior to force clearing the TPM of these values and compliance state. Another option would be to keep a TPM around that will always remain in the compliance state and can be referred to from time to time during application development.

One other thing: after you have leveraged the TPM_OIAP command to ascertain the compliance state of the TPM, it would be wise to power cycle the TPM or execute the command TPM_Terminate_Handle. The input message regarding this command is defined in Figure 11.3.

Authorization Tag	0x00	0xC1		
Parameter Size	0x00	0x00	0x00	0x0E
Ordinal	0x00	0x00	0x00	0x96
Authorization Handle	0x00	0x00	0x00	0x00

Figure 11.3 Determining compliance vector inclusion – TPM_Terminate_Handle input message

The TPM_Terminate_Handle, by the way uses the authorization handle defined in the TPM_OIAP output command message block, frees up the TPM resources consumed by the TPM_OIAP command. The output command message regarding this command is defined in Figure 11.4.

Authorization Tag	0x00	0xC4		
Parameter Size	0x00	0x00	0x00	0x0A
Return Code	0x00	0x00	0x00	0x00

Figure 11.4 Determining compliance vector inclusion – TPM_Terminate_Handle output message

11.2 TPM Initialization Regarding Compliance State

Before executing higher-order TPM commands, you must first perform a basic initialization sequence; note that this sequence pertains to compliance state only. The first order of business is to perform a TPM_StartUp command and, within this discussion, the only mode that will be described is the clear mode. The TPM_StartUp will initialize the TPM and perform a series of internal self-test functions. Figure 11.5 defines the input message block that invokes the TPM_StartUp command.

Authorization Tag	0x00	0xC1		
Parameter Size	0x00	0x00	0x00	0x0C
Ordinal	0x00	0x00	0x00	0x99
Startup Mode	0x00	0x01		

Figure 11.5 The TPM_StartUp, compliance mode, input message block

Authorization Tag	0x00	0xC4		
Parameter Size	0x00	0x00	0x00	0x0A
Return Code	0x00	0x00	0x00	0x00

Figure 11.6 The TPM_StartUp, compliance mode, output message block

After the TPM executes this command, the corresponding output message block, if successful, is defined in Figure 11.6. Depending on the version of TPM and the vendor, there may be an additional self-test-based command that may be required to execute within the TPM.

For the sake of simplicity, this book recommends the TPM_Continue SelfTest command execution, which places the TPM in a post-initialization state. If the TPM requires the execution of the TPM_Continue-SelfTest and this command is not supplied to the TPM after executing the TPM_StartUp command, the Atmel TPM will respond with a TCG_RETRY error. What this states is that the ContinueSelfTest command has been executed in place of the command sent to the TPM and that the TCG Software Stack (TSS) must resend the command it was trying to execute. Note that some Atmel TPMs do not require the ContinueSelfTest for a partial self-test execution, TPM_StartUp, and a continuation of the internal self-test – TPM_Continue SelfTest. To determine the actual requirements, please ask your TPM vendor, but it can never hurt to execute the TPM_ContinueSelfTest regardless.

With that said, let's look at the TPM_ContinueSelfTest command input message block described in Figure 11.7. This is a very simple command with

Authorization Tag	0x00	0xC1		
Parameter Size	0x00	0x00	0x00	0x0A
Ordinal	0x00	0x00	0x00	0x53

Figure 11.7 The TPM_ContinueSelfTest, compliance mode, input message block

a very simple passed or failed command response, with the typical output message block, as defined in Figure 11.8.

Authorization Tag	0x00	0xC4		
Parameter Size	0x00	0x00	0x00	0x0A
Return Code	0x00	0x00	0x00	0x00

Figure 11.8 The TPM_ContinueSelfTest, compliance mode, output message block

Once the TPM is in the post-initialization state and the deactivated or disabled states are not activated, the continuation of compliance-based TPM command execution can begin. Note that the TPM should never be deactivated or disabled unless you have taken a "compliance-based" TPM and force cleared this device. Chapter 23, TPM System Deployment Initialization, addresses the issue of recovering from a disabled or deactivated TPM state.

11.3 The Compliance Endorsement Key

The last issue that I would like to discuss with regard to the TPM compliance state is the notion of a compliance Endorsement Key (EK) – more specifically, the compliance EK public key. It goes without saying, apparently not, that you do not want to ship embedded systems with a TPM whose EK is based on predictable compliance data; but to facilitate the notion of low entropy, this public key must be well known. Therefore the next thing is to validate that the TPM EK public key is indeed a compliance-defined value, much like the TPM_OIAP command was used to determine TPM compliance state. The specific command details regarding the TPM_ReadPubek are discussed in Chapter 13, Establishing a TPM Owner, and focuses specifically on the public key returned from this command execution.

First, the prerequisite concerning TPM_ReadPubek is that the initialization sequence, defined earlier in this chapter, has been established. Next, command the TPM_ReadPubek and get the TPM response, noting the public key data. If the EK public key is identical to the public key defined in Figure 11.9, you're in compliance state and this public key will generate predictable command execution responses regarding future command execution dependent on this data – for example, the TPM_TakeOwnership command execution.

If the EK public key is not defined in Figure 11.9, then you are not in compliance mode and any attempt to correlate command execution with

Endorsement Compliance Public Key Data															
0xab	0x56	0x7c	0x0e	0x60	0x8c	0x5c	0x18	0x9e	0x90	0x2c	0x37	0x32	0xcf	0xe3	0xfe
0x4f	0xa7	0xb5	0x0c	0x78	0xa1	0x5d	0xa7	0x39	0xeb	0xc0	0x06	0x87	0x05	0xdb	0x1f
0xe4	0xab	0x2a	0x9a	0x68	0xe3	0x5b	0xb6	0xfb	0x27	0x69	0x5a	0x4b	0xe2	0x90	0x65
0x04	0xb2	0x78	0xcf	0x44	0x02	0x7c	0x16	0x4c	0xfb	0xf5	0xf0	0xf6	0x25	0x7d	0x31
0xf1	0x2e	0xd8	0x67	0x93	0x5a	0x48	0xb2	0xc1	0x4c	0x16	0xfd	0x97	0xe5	0x86	0x65
0x4a	0x2e	0x07	0x4b	0x14	0x78	0xf7	0x66	0x83	0x66	0x05	0xb0	0xea	0xec	0x1e	0x16
0xcf	0xf9	0xf9	0xc5	0x5c	0xbc	0x7b	0x42	0x24	0xa1	0xa7	0x1b	0x55	0xd7	0x4b	0xb1
0x62	0x7f	0x90	0x88	0xee	0xfb	0xfb	0x26	0xb1	0x4f	0x56	0x97	0x8c	0xd0	0x12	0x05
0xa6	0xef	0x09	0xc9	0x08	0x10	0xf2	0x1b	0x65	0x9c	0xf2	0x05	0x7b	0xcc	0x4e	0x6a
0x65	0x0c	0x1c	0xe1	0xb5	0x3e	0x86	0x7d	0xf8	0x0b	0x8b	0x6f	0xe3	0x72	0x2b	0xcb
0xc9	0x3d	0xf8	0x61	0xf4	0x83	0x74	0xb1	0x38	0xa6	0xce	0xde	0x18	0x7f	0x8d	0xc4
0x8f	0xa1	0x8e	0xa6	0xac	0x71	0xa4	0x89	0x60	0xd3	0x3e	0x5f	0x3d	0x18	0x5c	0x32
0x6c	0x96	0x1d	0x84	0x8b	0x50	0xc3	0x5b	0x68	0x5c	0x16	0x2d	0x9c	0xbb	0xf1	0x79
0x60	0x6e	0xc9	0x25	0xaa	0xec	0x26	0x9e	0x9e	0xd4	0xd6	0x89	0xf3	0xff	0x23	0xaa
0x75	0x46	0x3b	0x4a	0xea	0x1d	0xe5	0x03	0xb9	0xac	0x6d	0xf8	0x2d	0x88	0xff	0x84
0x12	0xb8	0x47	0xcf	0x3a	0x32	0xc9	0x66	0xc6	0xe3	0x2c	0x1f	0x7d	0x30	0xd8	0x99

Figure 11.9 EK public key data, compliance mode

compliance vectors will fail. Note that if you are confident your TPM is in compliance state, you can just use the defined public key described in the compliance vectors. The procedure of reading the EK public key is merely a fail-safe with regard to the TPM and compliance state.

In summary, the TPM must be initialized prior to executing the full command suite available in this device. The TPM_OIAP command can be leveraged to determine whether the TPM is in compliance state and contains the TCG-defined compliance vectors. If the TPM_OIAP is used to determine the compliance state, the TPM must be power cycled or the TPM_Terminate_ Handle must be executed within the device. The value associated with the TPM EK public key will "certify" that the compliance vectors are established within the device. *Certify*, in this context, loosely means that, depending on the value, the EK public key is indeed a compliance key and that the TPM is in compliance mode. In other words, if the TPM_ReadPubek returns the compliance key associated with the EK public key, then the TPM is in compliance mode and future TPM command execution, dependent on the EK public key, will produce predictable results. This method of determining TPM compliance state can be substituted for the TPM_OIAP-based test. In addition, the chapters devoted to command sequence examples will leverage the compliance mode and the EK public key compliance vector. With that said, let's take a look at some of the compliance vectors – the theme of the next chapter.

12

Compliance Vectors and Their Purpose

This chapter covers the compliance vectors, which define a state of low entropy and allow predictable results concerning Trusted Platform Module (TPM) command execution. Without these compliance vectors, defined in the Trusted Computing Group (TCG) Compliance Configuration Specification, it would be virtually impossible for TPM vendors to prove compliance regarding the TCG Main Specification. In addition to compliance testing, the compliance vectors allow end users to develop applications that leverage the predictability provided by them and allow me to give examples that can be repeated to leverage predictable results.

The concept regarding compliance vectors is to produce consistent, non-random-based RSA keys, random number generation, platform configuration register (PCR) digests, and nonce values in an effort to promote predictability. Eventually, the embedded systems developer will want to purge these vectors from the TPM and change the TPM state to use nonpredictable high-entropy-based cryptographic and random data. The mechanism for such a TPM state change is to simply leverage the TPM_ForceClear command, which will delete all compliance material from the TPM and enable random behavior. With that said, let's look at the compliance vectors as defined within specific categories.

12.1 The Compliance RSA Keying Material

The TPM must be able to leverage well-known data to facilitate predictable results regarding TCG compliance testing. One basic foundation for this result concerns the RSA cryptographic keying material used in the TPM compliance state. The TCG Compliance Configuration Specification addresses these issues and this book focuses on the compliance data needed to reproduce the examples detailed in later chapters. For complete information regarding the compliance state for the TPM, refer the TCG Compliance Configuration Specification on the TCG's web site.

The first defined RSA key pair – public key in this book – is **Key A**. This key is used to describe the compliance endorsement key (EK) along with any additional key(s) generated by the user outside of the defined RSA key set existing within the TPM in compliance state. The public key concerning **Key A** is defined in Figure 12.1.

0xab	0x56	0x7c	0x0e	0x60	0x8c	0x5c	0x18	0x9e	0x90	0x2c	0x37	0x32	0xcf	0xe3	0xfe
0x4f	0xa7	0xb5	0x0c	0x78	0xa1	0x5d	0xa7	0x39	0xeb	0xc0	0x06	0x87	0x05	0xdb	0x1f
0xe4	0xab	0x2a	0x9a	0x68	0xe3	0x5b	0xb6	0xfb	0x27	0x69	0x5a	0x4b	0xe2	0x90	0x65
0x04	0xb2	0x78	0xcf	0x44	0x02	0x7c	0x16	0x4c	0xfb	0xf5	0xf0	0xf6	0x25	0x7d	0x31
0xf1	0x2e	0xd8	0x67	0x93	0x5a	0x48	0xb2	0xc1	0x4c	0x16	0xfd	0x97	0xe5	0x86	0x65
0x4a	0x2e	0x07	0x4b	0x14	0x78	0xf7	0x66	0x83	0x66	0x05	0xb0	0xea	0xec	0x1e	0x16
0xcf	0xf9	0xf9	0xc5	0x5c	0xbc	0x7b	0x42	0x24	0xa1	0xa7	0x1b	0x55	0xd7	0x4b	0xb1
0x62	0x7f	0x90	0x88	0xee	0xfb	0xfb	0x26	0xb1	0x4f	0x56	0x97	0x8c	0xd0	0x12	0x05
0xa6	0xef	0x09	0xc9	0x08	0x10	0xf2	0x1b	0x65	0x9c	0xf2	0x05	0x7b	0xcc	0x4e	0x6a
0x65	0x0c	0x1c	0xe1	0xb5	0x3e	0x86	0x7d	0xf8	0x0b	0x8b	0x6f	0xe3	0x72	0x2b	0xcb
0xc9	0x3d	0xf8	0x61	0xf4	0x83	0x74	0xb1	0x38	0xa6	0xce	0xde	0x18	0x7f	0x8d	0xc4
0x8f	0xa1	0x8e	0xa6	0xac	0x71	0xa4	0x89	0x60	0xd3	0x3e	0x5f	0x3d	0x18	0x5c	0x32
0x6c	0x96	0x1d	0x84	0x8b	0x50	0xc3	0x5b	0x68	0x5c	0x16	0x2d	0x9c	0xbb	0xf1	0x79
0x60	0x6e	0xc9	0x25	0xaa	0xec	0x26	0x9e	0x9e	0xd4	0xd6	0x89	0xf3	0xff	0x23	0xaa
0x75	0x46	0x3b	0x4a	0xea	0x1d	0xe5	0x03	0xb9	0xac	0x6d	0xf8	0x2d	0x88	0xff	0x84
0x12	0xb8	0x47	0xcf	0x3a	0x32	0xc9	0x66	0xc6	0xe3	0x2c	0x1f	0x7d	0x30	0xd8	0x99

Figure 12.1 The compliance Key A public key data

Therefore if you wanted, for example, to perform a TPM_TakeOwnership command while the TPM was in compliance state, this command requires the owner and Storage Root Key (SRK) secrets to be encrypted with the EK public key. You would simply encrypt these two secrets using the compliance data representing the EK public key. Since the corresponding compliance EK private key exists within the TPM, the decryption of the two secrets would be facilitated and the TPM_TakeOwnership command would pass. One note: there is additional compliance data that allows this command to predictably execute within the TPM; we will get to these issues later in the chapter.

Let's focus on the TPM_TakeOwnership command while discussing the next fixed RSA key pair defined by the compliance specification. This fixed key pair is identified as **Key B** and is summarized in Figure 12.2. This key represents the SRK RSA key pair that will be generated as a result of the successful execution concerning the TPM_TakeOwnership command.

0xe2	0xf1	0xd9	0xe8	0x77	0xf6	0xf5	0x7f	0x0b	0xd0	0x08	0x9e	0xba	0x37	0x37	0xc8
0x31	0x01	0xd1	0x0d	0x20	0xb7	0x98	0xdd	0x26	0x91	0xf1	0xa1	0x5a	0xb5	0x31	0xc7
0x11	0x86	0x71	0x95	0xf9	0x45	0x79	0x27	0x5a	0x5a	0xfb	0xa1	0x1c	0x3b	0x11	0x5f
0x07	0x8a	0x59	0x53	0xe8	0xb6	0x67	0xbd	0x84	0x1d	0x9c	0xf1	0xe5	0xcd	0x71	0x51
0xdd	0x9b	0x67	0xa7	0xd5	0x8d	0x3b	0x8a	0xe9	0x16	0xdf	0x93	0x92	0x1f	0x7d	0xbe
0xd9	0xab	0xf8	0x79	0x20	0x2a	0x29	0x0e	0x7d	0xf6	0x5b	0x71	0xd5	0xb2	0x6c	0x94
0x6b	0x1e	0xfc	0x09	0x66	0x4f	0x8b	0x7c	0x0d	0x68	0x32	0x0e	0xe9	0xe7	0xcc	0xa0
0x68	0xd0	0xc1	0x7e	0x4a	0xaf	0x52	0xa5	0x4e	0x9e	0x16	0x34	0x1a	0x1a	0x6c	0x44
0x40	0x8e	0xec	0x67	0xfc	0xd5	0x49	0x1e	0xc7	0x78	0x63	0xa2	0x68	0x11	0xe2	0x3e
0xe1	0x12	0x6b	0x80	0x9f	0xaf	0x88	0x21	0xfd	0x5f	0x66	0xfd	0x12	0x68	0xb4	0x1f
0x67	0xec	0x15	0x6e	0xb1	0xa4	0x2e	0x29	0x40	0xdc	0x5a	0xd8	0xab	0xa1	0xbb	0x5f
0x75	0x28	0x69	0x8f	0x03	0xe2	0xb7	0x7f	0x44	0x70	0x3a	0xbc	0x6d	0x75	0xf4	0x10
0xc8	0xe8	0x75	0x48	0xdf	0x84	0xe8	0x0f	0x46	0xa9	0x1b	0x3d	0x8f	0x98	0x68	0x64
0x48	0xb1	0x98	0x7e	0x85	0x1c	0x84	0xe1	0x92	0x44	0xae	0x7d	0x79	0xab	0x88	0x1e
0x79	0xe9	0x88	0xc7	0x32	0x51	0x37	0x74	0xa8	0x99	0xee	0x69	0xfb	0xbf	0x61	0xfd
0x0a	0xb5	0x82	0xc6	0x19	0x53	0x36	0x44	0x9a	0xf9	0xa1	0x8f	0x3a	0xca	0xdb	0x9d

Figure 12.2 The compliance Key B public key data

This book only focuses on the public key portion, since the private key is only known to the TPM and never is disclosed outside of this secure boundary. This compliance public key is normally used as a parent key when additional keying material is being generated or loaded into the TPM – besides the keys defined by the Compliance Configuration Specification. For example, when loading in externally generated RSA keys, the private data with regard to this key must be encrypted by a public key whose private key exists within the TPM. The compliance SRK public key would be leveraged during the encryption of the child key's private data when the SRK in indexed as the parent key.

The final fixed RSA key pair concerns the predefined keying material that exists in compliance form within the TPM. These keys consist of five RSA key pairs defined for specific purposes; two storage keys, a signing key, an identity key, and a binding key, in this order. The key indexing for these keys is 0x00000000, 0x00000001, 0x00000002, 0x00000003, and 0x00000004. The fixed RSA public key is identified as **Key C**, as shown in Figure 12.3.

These keys are used for specific TPM command verification and can be leveraged during application development if their use is defined by the specific command considered for execution. If not, additional RSA keys can be

0xbf	0x27	0xb1	0x3b	0xd0	0xf5	0xa4	0x97	0x41	0x3c	0x7e	0x5d	0x9e	0x6a	0xe4	0x73
0xc5	0xb7	0x7a	0x49	0xd3	0xa7	0xff	0xda	0xf2	0x3b	0x6e	0x95	0x69	0xc0	0x8e	0x40
0x5f	0xdf	0xd2	0xce	0xc3	0xb9	0xd4	0x6a	0x0e	0xc1	0x4d	0x54	0x17	0x8d	0xfd	0x63
0x38	0xea	0x28	0xaf	0x31	0xfc	0x22	0x2a	0x27	0x28	0x56	0xd1	0x6f	0xd3	0x3b	0xa0
0x46	0x05	0x65	0x9a	0x31	0x62	0xf1	0x14	0xae	0x1f	0x3a	0x81	0xd0	0x31	0xa3	0xcf
0xae	0xf5	0x24	0xb4	0xa1	0x7c	0x9a	0x9b	0x3c	0xef	0x6c	0x98	0x3a	0xfb	0x6f	0xbf
0x58	0x03	0x05	0x58	0x49	0x63	0xa3	0xb0	0x7d	0xd2	0x39	0xe8	0x15	0xf6	0xf3	0x40
0xee	0x65	0x6b	0x10	0x7a	0xfc	0x43	0x4c	0x24	0x96	0x32	0x63	0x37	0xd0	0xe5	0xc6
0xc5	0x05	0x43	0xb8	0xca	0x35	0x30	0x22	0x61	0xa3	0x36	0x8d	0x29	0x35	0xa9	0x49
0x8e	0x82	0x6e	0x03	0x01	0xc3	0x99	0x93	0x8b	0x1b	0x07	0xdc	0xb4	0x73	0xc4	0x9e
0x59	0x32	0x5b	0x2a	0xbd	0x7b	0x54	0x21	0x4e	0x56	0x7a	0xca	0x70	0xa0	0x18	0xb1
0xf5	0x0f	0x57	0x81	0xd7	0x18	0x92	0xd7	0x73	0x31	0x4a	0x83	0x7f	0xc1	0x0b	0x0a
0x6e	0xa7	0x1d	0xcf	0x53	0x48	0xff	0x11	0xbe	0x01	0x96	0xdf	0x68	0x5b	0x63	0x1a
0x79	0x43	0xfe	0x3b	0x94	0xf4	0xdb	0x5e	0xf8	0xd8	0xe9	0xb0	0xf7	0x39	0x32	0x4e
0xc8	0x2e	0x0a	0x4f	0xbe	0x19	0xd8	0x6d	0x61	0x70	0xc0	0xa8	0x08	0xfc	0xc2	0x4f
0xa5	0x86	0x03	0x6a	0x0f	0x67	0x51	0x9c	0x1b	0x49	0x7a	0x54	0xe7	0x68	0x1f	0xeb

Figure 12.3 The compliance Key C public key data

loaded into the TPM via the SRK or the storage keys indexed by 0x00000000 and 0x00000001. Any key created in this manner will be defined by the compliance keying material labeled as **Key A**. Note that there are two versions of the compliance specification – the specification associated with the Version 1.1b Main Specification and another associated with the Version 1.2 Main Specification. The 1.2 Compliance Configuration Specification augments 1.1b and contains additional definitions regarding entities specific to Version 1.2.

12.2   The Compliance Nonces, Secrets, and Random Numbers

There are five different fixed 20-byte nonce values defined in the Compliance Configuration Specification; they are shown in Figure 12.4. These nonces are used at various times for fixed-nonce generation. For example, the nonceEven value generated by the TCG Software Stack (TSS) during command message input is defined as Nonce C. The corresponding nonceOdd, generated by the TPM in response to the input command message, will always be the Nonce Fixed 20-byte value. The other nonce values are used for the nonceOdd Object Specific Authorization Protocol (OSAP) and other intermediate conformance data, which are discussed in the chapters that deal with command sequencing examples.

Nonce A	0x88	0xbd	0x32	0x9a	0xf3	0x38	0xf9	0xf2	0xc9	0xef
	0xd7	0x7b	0x00	0x6c	0x41	0x98	0xab	0xc8	0x0c	0xf5
Nonce B	0xe3	0xd1	0x18	0x27	0x6e	0xcd	0x4f	0x47	0x87	0x77
	0xad	0x7c	0xd3	0x1f	0xfb	0xa6	0x50	0x5f	0x8a	0x40
Nonce C	0xb9	0x73	0x05	0xfa	0xdb	0xe3	0x4d	0xc5	0x46	0x65
	0x10	0x00	0x0a	0x55	0x04	0x2e	0x3f	0xea	0xbf	0x27
Nonce D	0xec	0x0c	0xc4	0xd3	0x26	0xed	0x5f	0xa5	0xf5	0xb1
	0x78	0x1b	0x37	0x5d	0x24	0xa3	0x36	0x36	0xe5	0x42
Nonce Fixed	0xA5	0xA5	0xA5	0xA5	0xA5	0xA5	0xA5	0xA5	0xA5	0xA5
	0xA5	0xA5	0xA5	0xA5	0xA5	0xA5	0xA5	0xA5	0xA5	0xA5

Figure 12.4 The five compliance nonce

In addition to the predefined nonces, 10 secrets are available concerning the different usage and entity authorization digest calculations. For example, the compliance specification defines the Usage Secret for the keying material indexed by 0x00000000 as Secret E and the Migration Secret for this same key as Secret J. The same is true for owner-authorized commands and the secret associated with this entity's authorization is Secret A. Figure 12.5 defines each secret's 20-byte value and the entity that these values are associated with. Furthermore, within the chapters discussing TPM command examples, some of these values will be leveraged to authorize command execution.

The TPM defined within the compliance state must leverage the ability concerning predictable random number generation. I know this sound's like a conflict in terms, but if a truly random value was produced then predictable results, with regard to command authorization and payload data, would not be possible. Therefore, any random data or padding values will

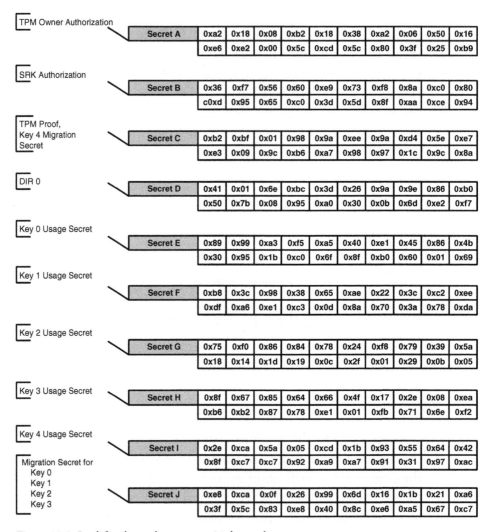

Figure 12.5 Predefined compliance secret 20-byte values

take form as a series of 0xA5, with a dependency on the length defined by the requested random number length or the padding size. Therefore, if the command TPM_GetRandom is executed with a requested length of

20 bytes, the command response will be 20 bytes of 0xA5s as defined by the compliance specification.

The other consideration regarding TPM compliance state is the issue of encoding, especially Optimal Asymmetric Encryption Padding (OAEP), which involves random data during message padding. For this subject, I refer readers to the Compliance Configuration Specification itself, which does a fine job of explaining the encoding procedures used in a TPM compliance state. One thing to keep in mind while reading this book is that this is not meant to be used outside of the TCG specifications and you should refer to them to help augment the basic content provided here. Another resource concerning the compliance state is the TPM vendor who must adhere to the requirements stated within the Compliance Configuration Specification.

12.3   The Compliance PCR Digest Values

The Platform Configuration Registers in TPM compliance state contain fixed 20-byte digest values that indicate a predictable host system configuration. These PCR digests are extendable via the TPM_Extend command, but caution must be used when issuing this type of command. For example, the keying material indexed by 0x00000002 is bound to PCRs 0, 1, and 3 or PcrA, PcrB, and PcrD. If either of these PCR digests is extended, any command leveraging this key will fail due to PCR modification issues.

The key that is indexed by 0x00000002 is an RSA signing key and therefore commands such as TPM-Sign and TPM_Quote, among others, will fail if any of the three listed PCR digests are extended. All of the other compliance keys are free from PCR binding and the other 13 PCRs may be extended, within compliance state, without affecting command execution due to referencing keying material that is bound to PCR(s). If you would like to include a RSA signing key that is not bound to any PCR(s) digest(s), you're free to create and load a signing key outside of the defined compliance keys.

For example, you could create a child key wrapped with the SRK facilitated by TPM_CreateWrapKey – that is, defined as a signing key – and load this keying material into the TPM via the TPM_LoadKey. If this is the first additional key, besides the compliance keys that are loaded into the TPM, a key handle of 0x00000005 will be assigned to this newly loaded key. Additional keying material can be loaded into the TPM as long as there are sufficient resources available; you might have to evict keying material that is not needed at the time to make room for the new key that is to be loaded.

Getting back to the fixed PCR digests: the Compliance Configuration Specification defines 16 PCR fixed 20-byte digests as defined in Figure 12.6.

PCR A	0xfd	0x89	0xa2	0xde	0x1a	0x91	0xd7	0xa2	0x2b	0xd1
	0x78	0x7a	0xa7	0xc2	0x77	0x9d	0xe0	0x99	0xf7	0xc0
PCR B	0x15	0x8f	0xd1	0x6a	0x35	0x8f	0x50	0x51	0x2a	0x81
	0x08	0xcf	0xe6	0xec	0xd0	0xf9	0x07	0xc5	0xc6	0x7c
PCR C	0x00	0xcd	0xd6	0xaf	0x54	0xc3	0x6f	0x3a	0x47	0x90
	0xc3	0xbc	0x5e	0xf9	0x31	0x3a	0xfa	0x73	0x26	0x03
PCR D	0xdc	0x62	0x90	0x0a	0x70	0x9b	0x29	0xcf	0x93	0xfa
	0x34	0x4c	0xd8	0xe6	0xc8	0x7e	0xc7	0xde	0xc5	0x35
PCR E	0x19	0xc0	0x4c	0xfb	0xf4	0x5e	0x38	0x92	0x6b	0xd6
	0x6d	0x64	0xdc	0xd1	0xf0	0xb8	0x30	0xb9	0x10	0x98
PCR F	0x8e	0x9e	0x63	0x27	0xa3	0x87	0x0b	0xe5	0xe7	0x41
	0xd6	0x7c	0x5d	0xc9	0x08	0x7d	0x7a	0x26	0xbe	0x00
PCR G	0x17	0x74	0x56	0x21	0xa9	0x45	0x7a	0x43	0x5c	0xad
	0x2e	0x9e	0x96	0x4c	0xee	0x6b	0x6c	0xec	0xfa	0x25
PCR H	0xe3	0x0d	0x10	0x07	0xe5	0x38	0x19	0x5d	0x25	0x1e
	0x8e	0x49	0x6e	0xde	0xbf	0x8f	0xae	0x38	0x20	0x21
PCR I	0xb9	0x1d	0x40	0x71	0xb0	0xab	0xaf	0x01	0xbd	0x14
	0x1d	0x2b	0x7c	0x5b	0xaf	0x66	0x9a	0xb7	0x2c	0x00
PCR J	0xd3	0xd4	0x51	0xb9	0xca	0x9d	0xfe	0x28	0xdc	0x5e
	0xad	0x02	0x9a	0x84	0x44	0x67	0x49	0x48	0x0a	0x87
PCR K	0x6a	0x30	0x46	0xf0	0x4e	0xdc	0xd3	0xa8	0xa5	0x4f
	0x4c	0x26	0x0f	0x64	0x63	0x0c	0x83	0x83	0xc7	0x3a
PCR L	0x42	0x5d	0x51	0x0a	0x0b	0x91	0x4c	0xa3	0x1f	0x76
	0x26	0x98	0xa8	0x97	0x8c	0x32	0x46	0xa0	0x92	0x6f
PCR M	0xbd	0x7d	0x9d	0x93	0xc7	0xb2	0x17	0x80	0x38	0xe3
	0x55	0xe9	0x45	0x19	0x3b	0x55	0x0a	0x3f	0xef	0x06
PCR N	0x39	0x0b	0x31	0x0a	0x42	0xec	0x07	0x07	0xa2	0x02
	0xe5	0xa6	0xd3	0xcb	0x8e	0xbb	0x33	0xfd	0x7c	0x0d
PCR O	0x98	0xbb	0x81	0x70	0xa6	0xf3	0x7b	0x3a	0x4b	0x79
	0x45	0xc0	0x15	0x2f	0xdc	0xee	0x5f	0xa1	0x1f	0x3b
PCR P	0x06	0x86	0x9d	0xe0	0xb9	0x0e	0x0e	0xd6	0x12	0x37
	0x5c	0x9c	0x68	0x74	0x67	0xd2	0x7e	0x47	0x7b	0xd4

Figure 12.6 The compliance PCR digests

These PCR digests will be leveraged during command execution examples later in this book.

The big issue to keep in mind regarding PCRs and compliance state has to do with the compliance key indexed by 0x00000002 and to remember that this key is tied to PCRs. If you modify the PCRs that are bound to this keying material, any command that references this entity will fail due to PCR configuration status. Other than this limitation, you are free to extend the remaining 13 PCRs defined by the Compliance Configuration Specification.

Establishing a TPM Owner

This chapter defines the commands that are directly related to the establishment of a Trusted Platform Module (TPM) Owner. Some of the commands described in this chapter are considered support-level commands, and without them, the establishment of a TPM Owner would not be possible. Therefore three TPM commands are defined in detail here: TPM_Create EndorsementKeyPair, TPM_ReadPubek, and TPM_TakeOwnership.

The first two are directly related to the third command in the sense that the Endorsement Key (EK) is used to encrypt the Owner and Storage Root Key (SRK) secrets during the process of establishing an owner via the TPM_ TakeOwnership command execution. TPM_CreateEndorsementKeyPair is included in this chapter because some TPM devices may not yet have a valid EK present in the TPM; this depends on the business relationship one has with the TPM vendor. If the EK is not present or a compliance EK is resident within the TPM, the TPM_CreateEndorsementKeyPair must be executed, hence its inclusion in this chapter.

13.1 The TPM_CreateEndorsementKeyPair Command

The establishment of an EK pair can only be executed once per TPM device and any attempts to execute this command after an EK has been established will result in a TCG_FAIL return code. The importance, regarding the EK, was established during the discussion about "root-of-trust" in Chapter 4. This chapter focuses on the cryptographic aspects of the EK with regard to taking ownership of the TPM along with the actual TPM command, which will establish the TPM Owner.

The first order of business is to determine whether the EK needs to be established within any given TPM device. The method for doing this is to first execute a TPM_ReadPubek and, based on the command execution results, proceed with either establishing the TPM Owner or a valid EK pair. Another means of performing this task is to execute the TPM_Create-EndorsementKeyPair, and based on this command's execution response, we can determine the next step that needs to be taken. Given that we are discussing the TPM_CreateEndorsementKeyPair, this command will be used and its execution results will be tested regarding how to proceed. If the command succeeds, we are done and can move right to the creation of a TPM Owner; if not, alternative solutions will be addressed. First let's look at the TPM_Create EndorsementKeyPair command specifics that allow this command to execute.

Figure 13.1 defines the command input message block regarding the individual parameters associated with the TPM_CreateEndorsementKeyPair. The first observation that can be made is in regard to the command authorization level – it is a non-authorized command. This makes sense because there are no

Authorization Tag	0x00	0xC1			
Parameter Size	0x00	0x00	0x00	0x36	
Ordinal	0x00	0x00	0x00	0x78	
Anti-replay nonce	AR1	AR2	AR3	AR4	AR5
	AR6	AR7	AR8	AR9	AR10
	AR11	AR12	AR13	AR14	AR15
	AR16	AR17	AR18	AR19	AR20
TCG_KEY_PARMS	0x00	0x00	0x00	0x01	
	0x00	0x03			
	0x00	0x01			
	0x00	0x00	0x00	0x0C	
	0x00	0x00	0x08	0x00	
	0x00	0x00	0x00	0x02	
	0x00	0x00	0x00	0x00	

Figure 13.1 TPM_CreateEndorsementKeyPair input message block

existing entities contained within the TPM prior to establishing the EK RSA key pair.

The next two parameters are of no surprise and they consist of the parameter size along with the command ordinal, which is defined in Figure 13.1. Next we come to the specific command payload, which consists of two parameters: an Anti-replay nonce and a Key Info Structure defined by TCG_KEY_PARMS. The Anti-replay nonce is an easy one; this value is no different than the nonceEven and nonceOdd values defined concerning authorization, a 20-byte random value. This value will be used in calculating a checksum regarding the newly created EK public key relative to the Anti-replay nonce, if the command execution is successful.

The Key Info Structure is of type TCG_KEY_PARMS and this Trusted Computing Group (TCG) structure is defined in Figure 13.2. Note, instead of defining all of the possible TCG-defined structures in a single chapter, this book describes each structure at the initial use within the context of command definition.

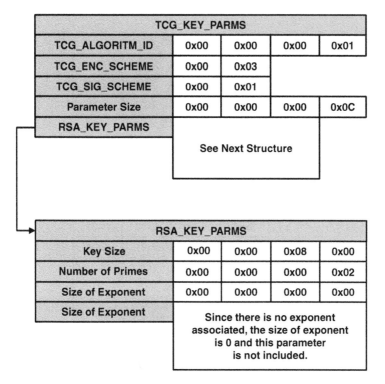

Figure 13.2 The TCG_KEY_PARMS structure

This structure, concerning the EK key type, defines the Algorithm ID as being RSA, an Encryption Scheme of TCG_ES_RSAESOAEP_SHA1_MGF1, a Signature Scheme of TCG_NONE, and a corresponding RSA-defined specific structure. The RSA-specific structure defines the RSA structure size as 12 bytes for a 2048-bit RSA key size, using two primes and no exponent defined. Thus the definition of the EK: an RSA key of at least 2048 bits is defined as an encryption key that doesn't allow RSA signatures to be produced using this key. The last step is to compile this command input message block and transmit it to the TPM via a Low Pin Count (LPC) bus or System Management Bus (SMBus).

The command output message block will be available after a short period of time; remember that the TPM is possibly generating a 2048-bit key pair in response to a successful command execution. If the command is successful, the TPM will respond with a command output message block as shown Figure 13.3.

The output message block contains the usual output header information in regard to the authorization tag, parameter size, and the return code. The

Authorization Tag	0x00	0xC4		
Parameter Size	0x00	0x00	0x01	0x3A
Ordinal	0x00	0x00	0x00	0x00
TCG_PUBKEY	0x00	0x00	0x00	0x01
	0x00	0x03		
	0x00	0x01		
	0x00	0x00	0x00	0x0C
	0x00	0x00	0x08	0x00
	0x00	0x00	0x00	0x02
	0x00	0x00	0x00	0x00
	0x00	0x00	0x01	0x00
	Endorsement Key's Public Portion			
Anti-replay nonce	AR1	AR2	AR3	AR4	AR5
	AR6	AR7	AR8	AR9	AR10
	AR11	AR12	AR13	AR14	AR15
	AR16	AR17	AR18	AR19	AR20

Figure 13.3 Successful TPM_CreateEndorsementKeyPair output message block

response payload comprises two parameters: the EK public key and the checksum, which is a Secure Hash Algorithm (SHA-1) hash of the EK public key concatenated with the Anti-replay nonce. Regarding the EK public key, this information is contained within the **TCG_PUBKEY** structure defined in Figure 13.4.

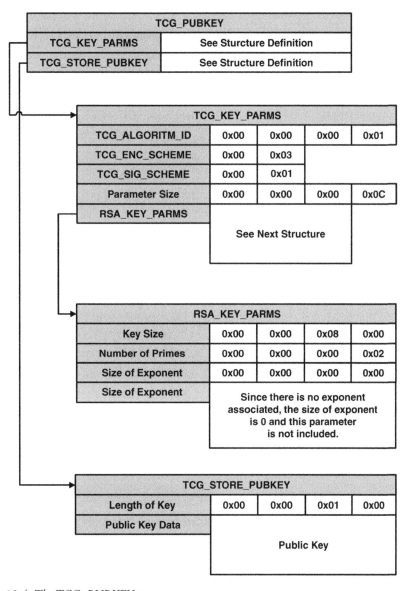

Figure 13.4 The TCG_PUBKEY structure

The Key Info, as described by the TCG_PUBKEY structure, contains the same information as defined by the TCG_KEY_PARMS structure used during command input, with the addition of two parameters. These parameters include the public key size along with the public key itself with regard to the newly generated EK. If this command is successful, you can move right on to the TPM_TakeOwnership command if you desire. If you want to wait before taking ownership, you might want to store the EK public key within non-volatile storage for use at a later time. This will save you the trouble of having to execute the TPM_ReadPubek to obtain the EK public key needed to support the execution of the TPM_TakeOwnership command.

OK, what happens if this command fails? Well this depends on the state of your TPM prior to executing the TPM_CreateEndorsementKeyPair, so let's investigate this a bit. Suppose that the TPM failed the TPM_Create EndorsementKeyPair execution with the command response TCG_FAIL. I know, very descriptive failure response. There are two possibilities that could be related to the EK that is more than likely resident within the TPM, which is why the command failed in the first place. What you have to do next is to read the EK public key by executing the TPM_ReadPubek command, and depending on this command response, you might be ready to establish a TPM Owner.

If you execute the TPM_ReadPubek and this command returns a public key, there are two possibilities: the EK is a valid randomly generated RSA key or the EK is a compliance key. If the EK is a compliance key, you can determine this by comparing the returned EK public key to the compliance EK public key defined in Compliance Vectors and Their Purpose, Chapter 12 of this book; you must force clear the TPM. See Chapter 23 regarding TPM initialization to determine the steps necessary for this task. After the TPM has been force cleared, this removes the compliance vectors from the TPM; you can now execute the TPM_CreateEndorsementKeyPair command; this should be successful on this go around. If the command still fails, check your input command message block concerning errors in the creation of these vectors.

13.2   The TPM_ReadPubek Command

This command is fairly straightforward and it involves commanding the TPM to return the EK public key. The command input message block for this command is listed in Figure 13.5. This message block is very similar to

Authorization Tag	0x00	0xC1			
Parameter Size	0x00	0x00	0x00	0x1E	
Ordinal	0x00	0x00	0x00	0x7C	
Anti-replay nonce	AR1	AR2	AR3	AR4	AR5
	AR6	AR7	AR8	AR9	AR10
	AR11	AR12	AR13	AR14	AR15
	AR16	AR17	AR18	AR19	AR20

Figure 13.5 TPM_ReadPubek input message block

the TPM_CreateEndorsementKeyPair and contains usual input header information; notice that this command is also a non-authorized command.

The only command payload parameter consists of the same Anti-replay nonce found within the command input message block for the TPM_ CreateEndorsementKeyPair command. Again the Anti-replay nonce is a 20-byte random number that will be used by the TPM to generate a checksum within the command's successful output message block. Compile the command input message block and transmit the data to the TPM. This command should execute quickly since we are only asking for the public key regarding an existing EK; otherwise, on command failure, the TPM responds very quickly.

The resulting successful command output message block is defined in Figure 13.6. To be brief, this command's output is identical in regard to the command output message block returned by the TPM_CreateEndorsement KeyPair. The checksum is calculated using the same algorithm just described with regard to the TPM_CreateEndorsementKeyPair output message and will differ because the Anti-replay nonce should differ, juxtaposing the two command input message blocks. Remember, the Anti-replay nonce, by definition, is a randomly generated value.

Now there are a few "got yeas" to watch out for. The most obvious failure, concerning this command, is that there is no EK public key inside the TPM to read and the TPM will respond with a failing return code of **TCG_NO_ ENDORSEMENT** if this is the case. One other possibility is that the TPM state is set so as not to allow the EK to be read, and if this is the case, the TPM will respond with a failing return code of **TCG_DISABLED_CMD**. If this command fails in some other way – for example, with a failing return code of **TCG_BAD_PARAM_SIZE** – check the command input message block for errors.

Authorization Tag	0x00	0xC4			
Parameter Size	0x00	0x00	0x01	0x3A	
Ordinal	0x00	0x00	0x00	0x00	
TCG_PUBKEY	0x00	0x00	0x00	0x01	
	0x00	0x03			
	0x00	0x01			
	0x00	0x00	0x00	0x0C	
	0x00	0x00	0x08	0x00	
	0x00	0x00	0x00	0x02	
	0x00	0x00	0x00	0x00	
	0x00	0x00	0x01	0x00	
	Endorsement Key's Public Portion				
Anti-replay nonce	AR1	AR2	AR3	AR4	AR5
	AR6	AR7	AR8	AR9	AR10
	AR11	AR12	AR13	AR14	AR15
	AR16	AR17	AR18	AR19	AR20

Figure 13.6 TPM_ReadPubek output message block

13.3   The TPM_TakeOwnership Command

Well, here we come to the mother of all TPM commands regarding authorization complexity. I say this because the TPM_TakeOwnership command has a strange method concerning authorization; it uses a deferred authorization protocol. For more information regarding this type of authorization protocol, see Chapter 10 on atypical authorization. Considering this command a little further, when you think about it, there is no parent key to facilitate a straightforward key load within the TPM regarding this command's execution; a TPM_LoadKey gets the child's key pair into the TPM by leveraging a parent key stored in the TPM. The problem is that there are no possible parent keys residing in the TPM concerning TPM state with regard to this command execution. So the TPM has to be creative by leveraging the EK – the only RSA keying material residing in the TPM at this moment – to facilitate the execution regarding the TPM_TakeOwnership command. One good note concerning security: by using the EK private key to encrypt the Owner and SRK Usage Secrets, the root-of-trust is cryptographically extended to each of these entities.

Now let's look at the input message block concerning this command, as described in Figure 13.7. Again, the usual header parameters, and notice that this command is a single authorized command; it is an Owner-authorized command. This is interesting, because there is not yet an Owner associated with the TPM during the time of execution; this command must authorize itself, so to speak, by executing a deferred authorization. *Deferred* meaning that the TPM authorization code cannot authorize this command until the Owner Secret, encrypted by the EK public key, is decrypted internally by the TPM. The other six parameters make up this command's payload and are defined in Figure 13.8.

The first payload parameter is of type TCG_PID_OWNER, which has a value defined as 0x0005. The TCG_PID_OWNER parameter is a subset of parameters defined by the TCG_PROTOCOL_ID and the possible values this type defines are listed in Figure 13.9.

The next four command input payload parameters are with regard to the EK public key encrypted values concerning the Owner Secret and the SRK Usage Secret, respectively. The first two parameters consist of the Owner Secret encrypted blob size (remember blob is crypto speak regarding a cipher) and the blob itself. In this case, the blob size is conveyed within 4 bytes, the EK is an RSA 2048-bit key, and the 256-byte blob, making 260 bytes in total. The next two parameters are identical except this blob contains the encrypted SRK Usage Secret and adds another 260 bytes to the input message block. The last parameter existing in the command input message block consists of a structure defined by the type TCG_KEY, which is defined in Figure 13.10.

The first parameter, regarding the TCG_KEY structure, is that of the TCG_VERSION and contains the major and minor versions as defined by this type (see Figure 13.11). This value identifies the major version – the version of the TCG specification the TPM was designed to – and the minor version – the Vendor firmware version existing within the device. Atmel uses both bytes to indicate the read-only memory (ROM) firmware version and the electrically

Authorization Tag	0x00	0xC2			
Parameter Size	0x00	0x00	0x02	0x70	
Ordinal	0x00	0x00	0x00	0x0D	
TCG_PROTOCOL ID	0x00	0x05			
Size of Enc Owner Secret	0x00	0x00	0x01	0x00	
Encrypted Owner Secret	Owner Sercret Encrypted by the EK Public Key				
Size of Enc SRK Secret	0x00	0x00	0x01	0x00	
Encrypted SRK Secret	SRK Sercret Encrypted by the EK Public Key				
TCG_KEY	0x01	0x01	0x00	0x06	
	0x00	0x11			
	0x00	0x00	0x00	0x00	
	0x00				
	0x00	0x00	0x00	0x00	
	0x00	0x03			
	0x00	0x01			
	0x00	0x00	0x00	0x0C	
	0x00	0x00	0x08	0x00	
	0x00	0x00	0x00	0x02	
	0x00	0x00	0x00	0x00	
	0x00	0x00	0x00	0x00	
	0x00	0x00	0x00	0x00	
	0x00	0x00	0x00	0x00	
Authorization Handle	0x00	0x00	0x00	0x00	
nonceOdd	NO1	NO2	NO3	NO4	NO5
	NO6	NO7	NO8	NO9	NO10
	NO11	NO12	NO13	NO14	NO15
	NO16	NO17	NO18	NO19	NO20
Continue Auth Session	0x00				
Authorization Digest	AD1	AD2	AD3	AD4	AD5
	AD6	AD7	AD8	AD9	AD10
	AD11	AD12	AD13	AD14	AD15
	AD16	AD17	AD18	AD19	AD20

Figure 13.7 TPM_TakeOwnership input message block

TCG_PROTOCOL ID	0x00	0x05		
Size of Enc Owner Secret	0x00	0x00	0x01	0x00
Encrypted Owner Secret	Owner Sercret Encrypted by the EK Public Key			
Size of Enc SRK Secret	0x00	0x00	0x01	0x00
Encrypted SRK Secret	SRK Sercret Encrypted by the EK Public Key			
TCG_KEY	0x01	0x01	0x00	0x06
	0x00	0x11		
	0x00	0x00	0x00	0x00
	0x00			
	0x00	0x00	0x00	0x00
	0x00	0x03		
	0x00	0x01		
	0x00	0x00	0x00	0x0C
	0x00	0x00	0x08	0x00
	0x00	0x00	0x00	0x02
	0x00	0x00	0x00	0x00
	0x00	0x00	0x00	0x00
	0x00	0x00	0x00	0x00
	0x00	0x00	0x00	0x00

Figure 13.8 TPM_TakeOwnership payload parameters

TCG_PROTOCOL ID	
TCG_PID_OIAP	0x0001
TCG_PID_OSAP	0x0002
TCG_PID_ADIP	0x0003
TCG_PID_ADCP	0x0004
TCG_PID_OWNER	0x0005

Figure 13.9 The TCG_PROTOCOL_ID type

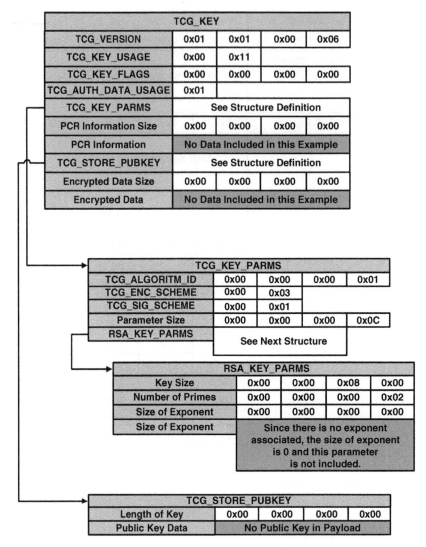

Figure 13.10 The TCG_KEY structure

TCG_VERSION	
Major	Version Byte, Usually the TCG Spec
Minor	Version Byte, Usually the TCG Spec
Revision Major	Byte Representing the FW Version
Revision Minor	Byte Representing the FW Version

Figure 13.11 The TCG_VERSION type

erasable programmable read-only memory (EEPROM) firmware version, respectively. The next parameter concerns the TCG type **TCG_KEY_USAGE** that, in this case, is defined to be of type **TCG_KEY_STORAGE**; see Figure 13.12 concerning the possible value for this type.

TCG_KEY_USAGE	
TPM_KEY_SIGNING	Signing Key
TPM_KEY_STORAGE	Storage Key
TPM_KEY_IDENTITY	Identity Key
TPM_KEY_AUTHCHANGE	Authorization Change Key
TPM_KEY_BIND	Binding Key
TPM_KEY_LEGACY	Legacy Key

Figure 13.12 The TCG_KEY_USAGE type

Now we come to the **TCG_KEY_FLAGS** type that adds additional attributes to the RSA key pair that is about to be created; in this case, the value is set to 0x00000000, which indicates that there is no modifier associated with the key about to be generated within the TPM. Figure 13.13 shows this TCG type.

TCG_KEY_FLAGS	
Redirection	0x00000001
Migration	0x00000002
Volatile	0x0000004

Figure 13.13 The TCG_KEY_FLAGS bit mask values

Moving on, the next parameter within the **TCG_KEY** structure concerns the **TCG_AUTH_DATA_USAGE** type, which is set to a value of 0x01 and indicates that the use of the SRK, about to be generated, must always be authorized. See Figure 13.14 for a description the TCG type.

Hey, the next parameter is a TCG structure we have seen before, the **TCG_KEY_PARMS** as defined in Figure 13.2 (see the TPM_CreateEndorsement

TCG_AUTH_DATA_USAGE	
Authorization Not Required	0x00
Authorization Required	0x01

Figure 13.14 The TCG_AUTH_DATA_USAGE type

KeyPair command). This parameter is set identically with regard to the input message block for the TPM_CreateEndorsementKeyPair command; the key about to be created is an RSA 2048-bit key size that uses two primes and has no exponent explicitly defined.

Here comes the tricky part regarding Platform Configuration Registers (PCRs) and the ability of a TPM key to be dependent on host system configuration. In other words, the use of the key that is about to be generated will be tied to the digest stored within one or more PCRs. This means that if the host system configuration changes, concerning any of the PCR(s) that this key references, the use of this key, within the TPM, will be prohibited. There are much better places to discuss the dependency of key usage concerning PCRs; the TakeOwnership command is complicated enough. For more information regarding this TPM capability, see Chapter 15 about the TPM_LoadKey. Note that the SRK about to be created, for this example, will not leverage this capability and therefore the next parameter, **PCRInfoSize,** will be set to a value of 0x00000000, which means that there will be no **PCRInfo** data in the input message block.

OK, only a few more parameters left: two to be exact. The next parameter concerns the **TCG_STORE_PUBKEY** structure as defined in Figure 13.15. There is no public key associated with the SRK; it hasn't been created yet, so the first value of this structure is set to 0x00000000, indicating that there is no data to be conveyed concerning this parameter.

TCG_STORE_PUBKEY				
Length of Key	LK1	LK2	LK3	LK4
Public Key Data	Public Key			

Figure 13.15 The TCG_STORE_PUBKEY structure

The last parameter, even though the TCG_KEY structure defines two more parameters, concerns the **encSize** parameter that will be set to 0x00000000. This indicates that there is no associated encrypted data within this input command message block. For edification purposes, the TCG_STORE_PUBKEY, **encSize** and **encData** are populated when commanding the TPM_LoadKey. Since we are generating a key, meaning these parameters have no meaning in this context, the parameters are defined as zero.

Finally, there is an attached authorization block containing the Owner-authorized authorization digest and associated authorization parameters (see Chapter 14 about TPM authorization). Note that the TCG Software Stack (TSS) uses the Owner Secret, encrypted with the EK public key, to authorize this command. The TPM will decrypt the Owner Secret to authorize the TPM_TakeOwnership command prior to execution within the TPM. The only thing left to do is to compile this input message block and transmit it to the TPM for execution. The resulting TPM output message block, with regard to the successful execution of the TPM_TakeOwnership command is defined in Figure 13.16.

Once again, the standard output header information: the authorization tag, defined as a single authorized output message, parameter size, and command return code. The only payload datum concerns a TCG_KEY structure that contains the SRK public key data. This structure is identical, regarding content, as defined within the corresponding command input message block with one exception; the structure now contains the SRK public key data. The reason for this pertains to the existence of the SRK within the TPM; one of the aspects of taking ownership of the TPM is to create an SRK. Notice that there is still no encryption data attached to the output message block as defined by the **encSize** and **encData** parameters. This is so because the SRK is defined as nonmigratable and thus the private key never leaves the TPM, even in encrypted form.

Attached to the output message block is the authorization block that allows the TSS to validate the authenticity of the command execution output. There are numerous error response codes associated with this command execution, one of which indicates the lack of an EK in the TPM – fairly difficult to decrypt using the EK if there is no EK present. This failure response is indicated by the return code TCG_NO_ENDORSEMENT. In addition, the TPM state can be such so as not to allow an Owner to be established, and if so, the error response will be TCG_INSTALL_DISABLED. If there is an error

Authorization Tag	0x00	0xC5		
Parameter Size	0x00	0x00	0x01	0x62
Return Code	0x00	0x00	0x00	0x00
TCG_KEY	0x01	0x01	0x00	0x06
	0x00	0x11		
	0x00	0x00	0x00	0x00
	0x00			
	0x00	0x00	0x00	0x00
	0x00	0x03		
	0x00	0x01		
	0x00	0x00	0x00	0x0C
	0x00	0x00	0x08	0x00
	0x00	0x00	0x00	0x02
	0x00	0x00	0x00	0x00
	0x00	0x00	0x00	0x00
	0x00	0x00	0x01	0x00
	SRK Public Key			
nonceEven	NE1	NE2	NE3	NE4	NE5
	NE6	NE7	NE8	NE9	NE10
	NE11	NE12	NE13	NE14	NE15
	NE16	NE17	NE18	NE19	NE20
Continue Auth Session	0x00			
Authorization Digest	AD1	AD2	AD3	AD4	AD5
	AD6	AD7	AD8	AD9	AD10
	AD11	AD12	AD13	AD14	AD15
	AD16	AD17	AD18	AD19	AD20

Figure 13.16 TPM_TakeOwnership output message block

regarding the key type to be created, check to make sure your algorithm ID and key attributes are set correctly within the input command message block; the return code will be TCG_BAD_KEY_PROPERTY. Finally, if the key usage is defined to anything other than TCG_KEY_STORAGE or the key flags have been set to migratable, the SRK cannot be migrated from the TPM; the resulting return code will be **TCG_INVALID_KEYUSAGE**. The best bet regarding any of these failures, with the exception of the **TCG_NO_ENDORSEMENT** error, is to check your command input message block for errors.

Now that we have defined a TPM Owner, let's see what we can do regarding TPM command execution relative to this entity. The next chapter defines the various TPM Command Suites that involve the Owner of the TPM.

Owner-Authorized Command Suite

This chapter deals with the command suite that establishes the set of Trusted Platform Module (TPM) Owner-authorized command execution. What this implies is that the Owner of the TPM can administer not only this entity itself, but also all the entities requesting the services of the TPM. So, going back to the information systems (IS) department, the Owner of the TPM within the context of a PC, can give the user limited access via keys created for this user and deny access to administrative TPM functionality. These types of administrative privileges could be the clearing of the TPM, access to the Endorsement Key (EK) public key, and access to the Storage Root Key (SRK) cryptographic capabilities deemed outside of the TPM user's scope.

These TPM administrative privileges are under the Owner's complete control, as defined by the numerous commands available to this entity and is the subject of this chapter. These commands are only available after the TPM Owner has been established (obvious statement), but again need to be written. With that, let's look at the suite of TPM commands available to the Owner of the device.

Before we do that, remember this book is concerned with the TPM basics as applied to embedded systems, so some TPM Owner commands will not be considered and their definition can be found within Version 1.1b of the TPM Main Specification itself. If you know of a TPM Owner-based command that is not described in this chapter or this book, refer to the Trusted Computing Group's (TCG) Main Specification for command details. The reason for this is to provide readers with the best information about the TPM concerning this device's inclusion within an embedded realization. Therefore the book, after defining the selected TPM commands to cover, can demonstrate examples of command sequencing, which produces overall system security.

14.1 The TPM_GetCapabilityOwner

This command differs from the more commonly used TPM_GetCapability; the command infers use by both the TPM Owner and User entities. What this command returns, to the TPM Owner, is the state of the TPM represented by a set of volatile and non-volatile flags; these flag structures will be described after command definition. With that, let's look at the input command message block, as defined in Figure 14.1.

Authorization Tag	0x00	0xC2			
Parameter Size	0x00	0x00	0x00	0x37	
Ordinal	0x00	0x00	0x00	0x66	
Authorization Handle	0x00	0x00	0x00	0x00	
nonceOdd	NO1	NO2	NO3	NO4	NO5
	NO6	NO7	NO8	NO9	NO10
	NO11	NO12	NO13	NO14	NO15
	NO16	NO17	NO18	NO19	NO20
Continue Auth Session	0x00				
Authorization Digest	AD1	AD2	AD3	AD4	AD5
	AD6	AD7	AD8	AD9	AD10
	AD11	AD12	AD13	AD14	AD15
	AD16	AD17	AD18	AD19	AD20

Figure 14.1 TPM_GetCapabilityOwner input message block

This is a very simple TPM capability request, juxtaposed against the TPM_GetCapability command, and has the header information with the Owner authorization block attached. In other words, this command requests the TPM's persistent and volatile state be reported to the Owner of the device. The command output message block is defined in Figure 14.2.

This command output message contains payload data concerning the TCG Version, the non-volatile flags, and the volatile flags. Each category of flag states is defined within 4 bytes, 8 bytes in total. Of course, the output message block has attached an authorization block that validates the Owner authorization command response. The TCG Software Stack (TSS) would simply recalculate the authorization digest, based on the Owner secret, to verify the authenticity of the message – it absolutely was generated by the TPM and is referencing the device owner.

Authorization Tag	0x00	0xC5			
Parameter Size	0x00	0x00	0x00	0x3F	
Return Code	0x00	0x00	0x00	0x00	
TCG Version	TV1	TV2	TV3	TV3	
Non-volatile Flags	NF1	NF2	NF3	NF4	
Volatile Flags	VF1	VF2	VF3	VF4	
nonceEven	NE1	NE2	NE3	NE4	NE5
	NE6	NE7	NE8	NE9	NE10
	NE11	NE12	NE13	NE14	NE15
	NE16	NE17	NE18	NE19	NE20
Continue Auth Session	0x00				
Authorization Digest	AD1	AD2	AD3	AD4	AD5
	AD6	AD7	AD8	AD9	AD10
	AD11	AD12	AD13	AD14	AD15
	AD16	AD17	AD18	AD19	AD20

Figure 14.2 TPM_GetCapabilityOwner output message block

The TCG Version information simply states the version of the TPM device being addressed. The non-volatile flag(s) are contained within 4 bytes and are defined in Figure 14.3.

The first flag, contained within the group of flags categorized as "non-volatile" or "persistent", is the **disable** flag, which signifies that the TPM is in a disabled state and limits command execution to the commands listed in Figure 14.4. Any attempted command execution that is outside of this "disabled state" list will return the error code of TCG_DISABLED. Therefore if you are integrating application(s) that leverage the TPM functionality and you get this response, the TPM is disabled and you will have to enable the TPM again to move forward (see Chapter 23 describing TPM initialization).

The **ownership** flag defines the ability to take ownership of the TPM device. If this flag is set to False (0), the TPM will not allow any attempts to take ownership of the device and will fail these attempts with the error code TPM_INSTALL_DISABLED. This flag does not state that there is an Owner established, but that TPM ownership is prohibited. The **deactivated** flag is used during the TPM_StartUp command execution and indicates how the TPM shall perform the Startup (clear); the TPM_StartUp (clear) will set the volatile flag to indicate that the TPM is in a deactivated state. Therefore, this flag governs the execution of the TPM_StartUp (clear) command, but has no

TCG PERSISTANT FLAGS	
disable	bit0
ownership	bit1
deactivated	bit2
readPubek	bit3
disableOwnerClear	bit4
allowMaintenance	bit5
physicalPresenceLifetimeLock	bit6
physicalPresenceHWEnable	bit7
physicalPresenceCMDEnable	bit8
CEKPUsed	bit9
TPMPost	bit10
TPMPostLock	bit11
Reserved	bit12 –bit15

Figure 14.3 The TCG PERSISTANT flag definition

DISABLED COMMAND SUITE
TPM_Reset
TPM_Init
TPM_StartUp
TPM_SaveState
TPM_SHA1Start
TPM_SHA1Update
TPM_SHA1Complete
TPM_SHA1CompleteExtend
TSC_PhysicalPresence
TPM_OIAP
TPM_OSAP
TPM_GetCapability
TPM_Extend
TPM_OwnerSetDisable
TPM_PhysicalEnable
TPM_ContinueSelfTest
TPM_SelfTestFull
TPM_GetTestResult
TPM_TerminateHandle

Figure 14.4 The TPM disabled command suite

direct control of the deactivated state regarding the command execution level. The **readPubek** flag is simply the privilege of TPM non-Owner entities to gain access to the public EK. If this flag is set to false, then no entity within the TPM, other than the Owner, will have access to the public EK data.

The **disableOwnerClear** flag is in regard to the Owner of the TPM having the privilege of being able to clear the TPM via the set of Owner-authorized commands – for example, TPM_OwnerClear. If the state of the flag is True (1), the Owner will be allowed to clear the TPM, else, any Owner-authorized commands, regarding clear functionality, will fail with a return code of TCG_CLEARED_DISABLED. The **TPMPost** and **TPMPostLock** involve the TPM self-test state and expand the internal self-testing to include additional requirements – for example, Federal Information Processing Standard (FIPS) level testing. Unless you are a complete masochist or are doing government-level realizations that require this type of testing, do not enable this state. The TCG conformance group is looking at this level of self-test state and negotiations have been ongoing regarding the functional specifics of this modified TPM state. My point is that the definition of this state may be modified, and I would recommend not leveraging this TPM behavior, if you must, contact your TPM vendor.

The last persistent flag definition that I will be discussing concerns the method that was used to populate the TPM EK. If the **CEKPUsed** is set to a True (1) state, then the EK was created via the TPM_CreateEndorsementKey-Pair; otherwise, this keying material was populated within the TPM using some other vendor's manufacturing process. Be very concerned if this flag is set to the False (0) state; you must know explicitly how this keying material was popu-lated within your device. The reason for this concerns the EK private key, secu-rity, and the EK public key, privacy.

This data is extremely sensitive and one method TPM vendors could use to populate this keying material is referred to as *squirting*; the EK is written into non-volatile memory at the vendor's manufacturing location. This may seem great from a production standpoint, but the issue becomes one of secu-rity and privacy, yours! Do you trust this vendor with your private and public keying material that is the root-of-trust, provided by the TPM and extended to your deployed system? If the answer is no or maybe, do yourself a favor and create your own EK by executing the TPM_CreateEndorsementKeyPair within your manufacturing process.

The remaining flags concern the Physical Presence functionality and are very much vendor-specific. Contact your TPM vendor with regard to implementation specifics about these flags and the Physical Presence realization the vendor has designed within the TPM of choice.

The volatile flag(s) are contained in the next 4 bytes and are defined in Figure 14.5. These flags are only interesting during the power-on state

TCG VOLATILE FLAGS	
deactivated	bit0
disableForceClear	bit1
physicalPresence	bit2
physicalPresenceLocked	bit3
postInitialization	bit4
Reserved	bit5–bit15

Figure 14.5 The TCG VOLATILE flag definition

concerning the TPM, hence the term *volatile*. The **deactivated** flag indicates the state of the non-volatile flag of the same name and is set or cleared by the execution of the TPM_StartUp (clear) command.

To set this flag explicitly, one would execute the command TPM_SetTempDeactivated (True) or to clear this flag TPM_SetTempDeactivated (False). Notice the "Temp" command characteristic, which indicates that this command is modifying a flag that exists in volatile memory of the TPM and any modification will be lost at TPM power-on reset (POR). The deactivated state will limit the command execution to those commands listed in Figure 14.6 and any other command execution attempts will generate a TCG_ DEACTIVATED return response.

The **disableForceClear** flag will govern the execution concerning the TPM_ForceClear command, which will prevent any entity, other than the TPM Owner, from clearing the device. The **postInitialize** flag indicates the execution of any remaining TPM self-test requirements and will be set to False, indicating that the requirements of post-initialization have been met. The remaining flags **physicalPresence** and **physicalPresenceLocked** is vendor-specific with regard to their implementation, and should be discussed with the TPM vendor relative to the specific vendor documentation.

The concept to remember regarding these flags is that they are used to make the TPM Owner aware of TPM state, not to control said state. There are commands that will cause the TPM state represented by these flags to be altered, which are both TCG-specified and vendor-specific. Some of the

DEACTIVATED COMMAND SUITE
TPM_Reset
TPM_Init
TPM_StartUp
TPM_SaveState
TPM_SHA1Start
TPM_SHA1Update
TPM_SHA1Complete
TPM_SHA1CompleteExtend
TSC_PhysicalPresence
TPM_OIAP
TPM_OSAP
TPM_GetCapability
TPM_TakeOwnership
TPM_OwnerSetDisable
TPM_PhysicalDisable
TPM_PhysicalEnable
TPM_PhysicalSetDeactivated
TPM_ContinueSelfTest
TPM_SelfTestFull
TPM_GetTestResult
TPM_TerminateHandle

Figure 14.6 The deactivated TPM command suite

TPM Owner-authorized command, which alter these flag(s) state, are discussed in this chapter.

14.2 The TPM_DisablePubekRead

This command disables the reading of the EK public key by any entity other than the TPM Owner, which can be accomplished by commanding the TPM_OwnerReadPubek. This also means that any entity trying to access the EK public key by commanding the TPM_ReadPubek will get an TCG_ DISABLED_CMD return code, indicating that this command has been disabled. This is a simple command from a command payload perspective, and it is also a TPM Owner-authorized command, as shown in Figure 14.7.

The command output response simply indicates that the command executed successfully or that there was a problem setting this state within the TPM persistent flags. The command output message block is defined in Figure 14.8.

Authorization Tag	0x00	0xC2			
Parameter Size	0x00	0x00	0x00	0x37	
Ordinal	0x00	0x00	0x00	0x7E	
Authorization Handle	0x00	0x00	0x00	0x00	
nonceOdd	NO1	NO2	NO3	NO4	NO5
	NO6	NO7	NO8	NO9	NO10
	NO11	NO12	NO13	NO14	NO15
	NO16	NO17	NO18	NO19	NO20
Continue Auth Session	0x00				
Authorization Digest	AD1	AD2	AD3	AD4	AD5
	AD6	AD7	AD8	AD9	AD10
	AD11	AD12	AD13	AD14	AD15
	AD16	AD17	AD18	AD19	AD20

Figure 14.7 TPM_DisablePubekRead input message block

Authorization Tag	0x00	0xC5			
Parameter Size	0x00	0x00	0x00	0x33	
Return Code	0x00	0x00	0x00	0x00	
nonceEven	NE1	NE2	NE3	NE4	NE5
	NE6	NE7	NE8	NE9	NE10
	NE11	NE12	NE13	NE14	NE15
	NE16	NE17	NE18	NE19	NE20
Continue Auth Session	0x00				
Authorization Digest	AD1	AD2	AD3	AD4	AD5
	AD6	AD7	AD8	AD9	AD10
	AD11	AD12	AD13	AD14	AD15
	AD16	AD17	AD18	AD19	AD20

Figure 14.8 TPM_DisablePubekRead output message block

14.3 The TPM_OwnerReadPubek

Once the EK public key's read capability has been disabled, the only method to obtain this data is via the TPM_OwnerREadPubek command. This limits the data knowledge to the Owner of the TPM and no other entity within the TPM can gain access to this data. There are a variety of reasons for doing this, from a security standpoint, and since there are no commands within the TPM command suite that leverage this data with regard to successful execution, limiting this data access to the TPM Owner is ideal for privacy issues. With that said, let's look at the input command massage block regarding this Owner command, as seen in Figure14.9.

Authorization Tag	0x00	0xC2			
Parameter Size	0x00	0x00	0x00	0x37	
Ordinal	0x00	0x00	0x00	0x7D	
Authorization Handle	0x00	0x00	0x00	0x00	
nonceOdd	NO1	NO2	NO3	NO4	NO5
	NO6	NO7	NO8	NO9	NO10
	NO11	NO12	NO13	NO14	NO15
	NO16	NO17	NO18	NO19	NO20
Continue Auth Session	0x00				
Authorization Digest	AD1	AD2	AD3	AD4	AD5
	AD6	AD7	AD8	AD9	AD10
	AD11	AD12	AD13	AD14	AD15
	AD16	AD17	AD18	AD19	AD20

Figure 14.9 TPM_OwnerReadPubek input message block

The input command message is very straight forward, header information and an Owner-authorization block. Juxtaposing the TPM_OwnerReadPubek with its sister command TPM_ReadPubek, there is one difference: there is an Anti-replay nonce included within the Owner version of this command. The output is very similar to the TPM_ReadPubek command, as detailed in Figure 14.10. The output command includes header information, with the return code, and a single command output payload consisting of a TCG_PUBKEY structure, identical to the command response regarding the TPM_ReadPubek output message. One difference that can be noted is that the TPM_Owner ReadPubek does not include any checksum data, which makes sense because

Authorization Tag	0x00	0xC5			
Parameter Size	0x00	0x00	0x01	0x4F	
Return Code	0x00	0x00	0x00	0x00	
TCG_PUBKEY	EK Public Key Contained within the TCG_PUBKEY Structure				
nonceEven	NE1	NE2	NE3	NE4	NE5
	NE6	NE7	NE8	NE9	NE10
	NE11	NE12	NE13	NE14	NE15
	NE16	NE17	NE18	NE19	NE20
Continue Auth Session	0x00				
Authorization Digest	AD1	AD2	AD3	AD4	AD5
	AD6	AD7	AD8	AD9	AD10
	AD11	AD12	AD13	AD14	AD15
	AD16	AD17	AD18	AD19	AD20

Figure 14.10 TPM_OwnerReadPubek output message block

there is no Anti-replay nonce associated with this command's input message block. In addition, the command output has an Owner-authorized block attached. The output message block is depicted in Figure 14.10.

This command is the only means of retrieving the EK public key data from the TPM after the Owner has disabled access to this information via the TPM_DisablePubekRead, which disables the TPM_ReadPubek command execution. A big reason for limiting the EK public key knowledge to the TPM Owner is in regard to privacy concerns; remember, the TPM_ReadPubek is a non-authorized TPM command. This implies that any entity, including entities outside the TPM, can gain this information unless the Owner of the TPM prevents this type of disclosure.

14.4 The TPM_OwnerClear

Just as the TPM supports the clearing of the device via "user" entities, the TPM also supports the clearing of the TPM with regard to the Owner. This command will clear the TPM of all established states or entities, including the SRK and all cryptographic material, except the EK; all ownership-related data sets all Platform Configuration Registers (PCRs) to their default values and sets the persistent flag(s) to their default values. Basically, this command "resets" the TPM to a state prior to any ownership state and, as a result, removes any "user" presence within the device. The input message concerning this command execution is defined in Figure 14.11.

Authorization Tag	0x00	0xC2			
Parameter Size	0x00	0x00	0x00	0x37	
Ordinal	0x00	0x00	0x00	0x5B	
Authorization Handle	0x00	0x00	0x00	0x00	
nonceOdd	NO1	NO2	NO3	NO4	NO5
	NO6	NO7	NO8	NO9	NO10
	NO11	NO12	NO13	NO14	NO15
	NO16	NO17	NO18	NO19	NO20
Continue Auth Session	0x00				
Authorization Digest	AD1	AD2	AD3	AD4	AD5
	AD6	AD7	AD8	AD9	AD10
	AD11	AD12	AD13	AD14	AD15
	AD16	AD17	AD18	AD19	AD20

Figure 14.11 TPM_OwnerClear input message block

This is a very simple command header and attached Owner-authorization block. The resulting output message block is equally simple with an output header and attached Owner-authorization block. The command execution result simply declares a successful command completion or a failure code. The output command message block, regarding the TPM_OwnerClear, is defined in Figure 14.12.

Authorization Tag	0x00	0xC5			
Parameter Size	0x00	0x00	0x00	0x33	
Return Code	0x00	0x00	0x00	0x00	
nonceEven	NE1	NE2	NE3	NE4	NE5
	NE6	NE7	NE8	NE9	NE10
	NE11	NE12	NE13	NE14	NE15
	NE16	NE17	NE18	NE19	NE20
Continue Auth Session	0x00				
Authorization Digest	AD1	AD2	AD3	AD4	AD5
	AD6	AD7	AD8	AD9	AD10
	AD11	AD12	AD13	AD14	AD15
	AD16	AD17	AD18	AD19	AD20

Figure 14.12 TPM_OwnerClear output message block

14.5 The TPM_DisableOwnerClear

This Owner-authorized command is very intuitive: disable the privilege, regarding the TPM Owner, of clearing the TPM of existing Owner – entities and state. Some of you may be scratching your head in wonder as to why this command would ever be leveraged within an actual system. The short answer is that the logical entity, in this case the TPM Owner defined by command execution, may not be the physical Owner and as such may not have the right to clear the TPM.

There are many methods of architecting this type of relationship, with regard to logical and physical TPM ownership, and the best source of obtaining the specific or recommended implementation details is the TPM vendor. Sorry for the "cop-out", but this type of situation requires the TPM definition of Physical Presence, which is defined by each TPM vendor, and if considered "protected" information is defined within a Non-Disclosure Agreement (NDA). Nonetheless, the command still stands and the input message block regarding the TPM_DisableOwnerClear is defined in Figure 14.13.

Authorization Tag	0x00	0xC2			
Parameter Size	0x00	0x00	0x00	0x37	
Ordinal	0x00	0x00	0x00	0x5C	
Authorization Handle	0x00	0x00	0x00	0x00	
nonceOdd	NO1	NO2	NO3	NO4	NO5
	NO6	NO7	NO8	NO9	NO10
	NO11	NO12	NO13	NO14	NO15
	NO16	NO17	NO18	NO19	NO20
Continue Auth Session	0x00				
Authorization Digest	AD1	AD2	AD3	AD4	AD5
	AD6	AD7	AD8	AD9	AD10
	AD11	AD12	AD13	AD14	AD15
	AD16	AD17	AD18	AD19	AD20

Figure 14.13 TPM_DisableOwnerClear input message block

The command input is a straightforward header with attached Owner-authorization block; the ordinal indicates that we wish to disable the clearing of the TPM by the device Owner. The output message block is equally simple and is defined in Figure 14.14.

Authorization Tag	0x00	0xC5			
Parameter Size	0x00	0x00	0x00	0x33	
Return Code	0x00	0x00	0x00	0x00	
TCG Version	TV1	TV2	TV3	TV3	
Non-volatile Flags	NF1	NF2	NF3	NF4	
Volatile Flags	VF1	VF2	VF3	VF4	
nonceEven	NE1	NE2	NE3	NE4	NE5
	NE6	NE7	NE8	NE9	NE10
	NE11	NE12	NE13	NE14	NE15
	NE16	NE17	NE18	NE19	NE20
Continue Auth Session	0x00				
Authorization Digest	AD1	AD2	AD3	AD4	AD5
	AD6	AD7	AD8	AD9	AD10
	AD11	AD12	AD13	AD14	AD15
	AD16	AD17	AD18	AD19	AD20

Figure 14.14 TPM_DisableOwnerClear output message block

14.6 The TPM_OwnerSetDisable

This command sets or clears the persistent flag **disable**, as described within the discussion regarding the TPM_GetCapabilityOwner. The input command message block concerning the command is defined in Figure 14.15.

The input message includes the usual header information and Owner-authorization block. In addition, the command has a single payload parameter, True (0x01); sets the persistent flag and False (0x00); and clears the persistent flag. The output of this command invocation is defined in Figure 14.16 and attests to the state of command execution – successful or some failure code.

14.7 The TPM_ChangeAuthOwner

This TPM command allows the modification of the Owner Secret or the SRK Usage Secret; both secrets are associated with the Owner of the TPM. The input command message leverages the use of an **EncAuth** and therefore this command must index an Object Specific Authorization Protocol (OSAP)

Authorization Tag	0x00	0xC2			
Parameter Size	0x00	0x00	0x00	0x38	
Ordinal	0x00	0x00	0x00	0x6E	
Disable State	0x00				
Authorization Handle	0x00	0x00	0x00	0x00	
nonceOdd	NO1	NO2	NO3	NO4	NO5
	NO6	NO7	NO8	NO9	NO10
	NO11	NO12	NO13	NO14	NO15
	NO16	NO17	NO18	NO19	NO20
Continue Auth Session	0x00				
Authorization Digest	AD1	AD2	AD3	AD4	AD5
	AD6	AD7	AD8	AD9	AD10
	AD11	AD12	AD13	AD14	AD15
	AD16	AD17	AD18	AD19	AD20

Figure 14.15 TPM_OwnerSetDisable input message block

Authorization Tag	0x00	0xC5			
Parameter Size	0x00	0x00	0x00	0x33	
Return Code	0x00	0x00	0x00	0x00	
nonceEven	NE1	NE2	NE3	NE4	NE5
	NE6	NE7	NE8	NE9	NE10
	NE11	NE12	NE13	NE14	NE15
	NE16	NE17	NE18	NE19	NE20
Continue Auth Session	0x00				
Authorization Digest	AD1	AD2	AD3	AD4	AD5
	AD6	AD7	AD8	AD9	AD10
	AD11	AD12	AD13	AD14	AD15
	AD16	AD17	AD18	AD19	AD20

Figure 14.16 TPM_OwnerSetDisable output message block

authorization session. There is also a very confusing issue regarding the output message authorization protocol that will be addressed during this command's output message block discussion. The command input message has the usual header information and the TCG tag indicates a single authorized command with the additional requirement that the authorization session be of the type OSAP. This command has three payload parameters: the

PROTOCOL ID		
TCG_PID_OIAP	0x00	0x01
TCG_PID_OSAP	0x00	0x02
TCG_PID_ADIP	0x00	0x03
TCG_PID_ADCP	0x00	0x04
TCG_PID_OWNER	0x00	0x05

Figure 14.17 The TCG_PROTOCOL_ID definitions

TCG_Protocol_ID, **TCG_EncAuth** and **TCG_Entity_Type**. The protocol ID type(s) are defined in Figure 14.17.

In the case of the TPM_ChangeAuthOwner, the protocol ID is defined as the type **TCG_PID_ADCP**, which indicates that the protocol being leveraged is of the type used during the modification of authorization data. In this case, the protocol ID reflects the modification of the Owner of SRK Usage Secret. The **EncAuth** parameters is a 20-byte value representing the new authorization secret that is encrypted by means as described in the chapter defining this type of cryptographic protection (see Chapter 10 about atypical authorization). Finally, the Entity Type, in regard to this command, can be two separate values: the **TCG_ET_OWNER** or the **TCG_ET_SRK**, with the Owner referring to Owner-authorized commands and the SRK referring to the SRK usage. Figure 14.18 defines the possible types that are referenced within this TCG type.

TCG_ENTITY_ TYPE		
TCG_ET_KEYHANDLE	0x00	0x01
TCG_ET_OWNER	0x00	0x02
TCG_ET_DATA	0x00	0x03
TCG_ET_SRK	0x00	0x04
TCG_ET_KEY	0x00	0x05

Figure 14.18 The TCG_ENTITY_TYPE values

This command is obviously Owner-authorized and defines what datum is to be modified along with the protocol, in this case **TCG_PID_ADCP**. The example TPM_ChangeAuthOwner is going to command that the Owner-authorization secret be modified as indicated by the **TCG_Entity_Type** **TCG_ET_OWNER**. Figure 14.19 shows the input message block concerning this command.

Authorization Tag	0x00	0xC2			
Parameter Size	0x00	0x00	0x00	0x4F	
Ordinal	0x00	0x00	0x00	0x10	
Protocol ID	0x00	0x04			
EncAuth	NO1	NO2	NO3	NO4	NO5
	NO6	NO7	NO8	NO9	NO10
	NO11	NO12	NO13	NO14	NO15
	NO16	NO17	NO18	NO19	NO20
Protocol ID	0x00	0x02			
Authorization Handle	0x00	0x00	0x00	0x00	
nonceOdd	NO1	NO2	NO3	NO4	NO5
	NO6	NO7	NO8	NO9	NO10
	NO11	NO12	NO13	NO14	NO15
	NO16	NO17	NO18	NO19	NO20
Continue Auth Session	0x00				
Authorization Digest	AD1	AD2	AD3	AD4	AD5
	AD6	AD7	AD8	AD9	AD10
	AD11	AD12	AD13	AD14	AD15
	AD16	AD17	AD18	AD19	AD20

Must index an OSAP authorization session

Figure 14.19 TPM_ChangeAuthOwner input message block

The output message is a very simple passed or failed command response and is authorized with the new Owner-authorization secret. There has been some great debate concerning which secret will be used to validate the output message of this command with some individuals recommending that the previous authorization secret be leveraged. This is a concern that might be brought to the attention of your TPM vendor concerning clarification on how the company has implemented this command. Per the TCG Main Specification Version 1.1b, the output authorization will leverage the new secret associated with the entity being addressed via the TCG_Entity_Type. With that, let's look at a typical output message that results from this command execution as defined in Figure 14.20.

14.8 The TPM_AuthorizeMigrationKey

The final Owner-authorized command that is addressed within the context of this book is in regard to key migration. To facilitate the migration of internal keying material from the TPM, a migration key authorization or

Authorization Tag	0x00	0xC5			
Parameter Size	0x00	0x00	0x00	0x33	
Return Code	0x00	0x00	0x00	0x00	
nonceEven	NE1	NE2	NE3	NE4	NE5
	NE6	NE7	NE8	NE9	NE10
	NE11	NE12	NE13	NE14	NE15
	NE16	NE17	NE18	NE19	NE20
Continue Auth Session	0x00				
Authorization Digest	AD1	AD2	AD3	AD4	AD5
	AD6	AD7	AD8	AD9	AD10
	AD11	AD12	AD13	AD14	AD15
	AD16	AD17	AD18	AD19	AD20

Figure 14.20 TPM_ChangeAuthOwner output message block

TCG_MIGRATIONKEYAUTH structure must be created. This type of structure is created by the TPM Owner and is used during by the migration command(s). By creating the **TCG_MIGRATIONKEYAUTH** structure, the TPM Owner does not have to be involved during the execution of the migration commands themselves. So the basic idea is to define a public key along with authorization that will facilitate the migration of TPM keying material at a time when migration-specific commands are leveraged. With that, let's look at the input message block in regard to the TPM_AuthorizeMigrationKey command, as defined in Figure 14.21.

The header information and the Owner-authorization block are the more obvious parameters. The command payload defines two parameters as **TCG_MIGRATE_SCHEME** and **TCG_PUBKEY**. The **TCG_PUBKEY** has been defined throughout this book and you can refer the specific chapters that discuss this TCG structure in detail. The TCG **TCG_MIGRATE_SCHEME** values are defined in Figure 14.22.

Note that the last value defined, **TCG_MS_MAINT**, defines a migration scheme that applies to TPM maintenance and this type of command(s) are vendor optional. Please check with your TPM vendor with regard to TPM maintenance support. The first migration scheme, **TPM_MS_MIGRATE**, can be used by all migration commands that do not leverage the "Re-Wrap" mode in reference to the TPM_CreateMigrationBlob command. This second migration scheme is defined as **TPM_MS_REWRAP**. The specifics regarding usage for

Authorization Tag	0x00	0xC2			
Parameter Size	0x00	0x00	0x01	0x55	
Ordinal	0x00	0x00	0x00	0x2B	
Migration Scheme	0x00	0x01			
TCG_PUBKEY	Public Key that Will Wrap the Private Data				
Authorization Handle	0x00	0x00	0x00	0x00	
nonceOdd	NO1	NO2	NO3	NO4	NO5
	NO6	NO7	NO8	NO9	NO10
	NO11	NO12	NO13	NO14	NO15
	NO16	NO17	NO18	NO19	NO20
Continue Auth Session	0x00				
Authorization Digest	AD1	AD2	AD3	AD4	AD5
	AD6	AD7	AD8	AD9	AD10
	AD11	AD12	AD13	AD14	AD15
	AD16	AD17	AD18	AD19	AD20

Figure 14.21 TPM_AuthorizeMigrationKey input message block

TCG MIGRATION SCHEMES		
TCG_MS_MIGRATE	0x00	0x01
TCG_MS_REWRAP	0x00	0x02
TCG_MS_MAINT	0x00	0x03

Figure 14.22 The TCG_MIGRATE_SCHEME values

the two migration schemes is discussed in Chapter 21, which is devoted to migration commands.

The output message block with regard to this command includes the header information along with the Owner-authorization block. In addition, there is one payload parameter associated with this command response and it involves the TCG_MIGRATIONKEYAUTH structure; this structure is defined in Figure 14.23.

The Owner-authorized creation of the TCG_MIGRATIONKEYAUTH structure will facilitate the migration of keying material and is an input parameter

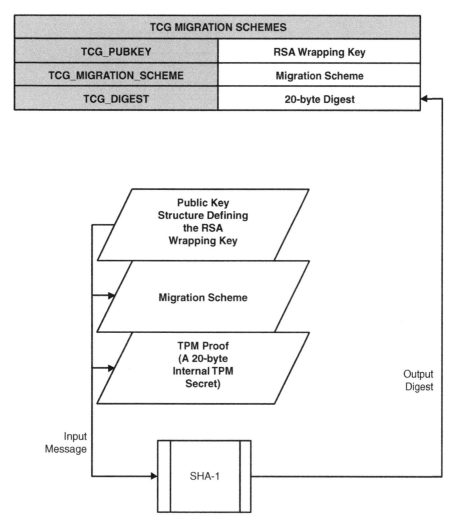

TCG MIGRATION SCHEMES	
TCG_PUBKEY	RSA Wrapping Key
TCG_MIGRATION_SCHEME	Migration Scheme
TCG_DIGEST	20-byte Digest

Figure 14.23 TCG_MIGRATIONKEYAUTH structure definition

to TPM_CreateMigrationBlob. The digest contained within the **TCG_ MIGRATIONKEYAUTH** structure is the Secure Hash Algorithm (SHA-1) digest of the message whose data is the concatenation of the migration key, migration scheme, and the 20-byte TPM Proof. The TPM Proof is a 20-byte value known only with the internal context of any given TPM; as such, the authorization facilitated by the **TCG_MIGRATIONKEYAUTH** structure pertains only to the TPM that created this structure. The complete description of the output message block, regarding the TPM_AuthorizeMigrationKey, is depicted in Figure 14.24.

Authorization Tag	0x00	0xC5			
Parameter Size	0x00	0x00	0x01	0x65	
Return Code	0x00	0x00	0x00	0x00	
TCG_MIGRATIONAUTH	Migration Authorization Structure that Can Be Used by the TPM_CreateMigrationBlob Command				
nonceEven	NE1	NE2	NE3	NE4	NE5
	NE6	NE7	NE8	NE9	NE10
	NE11	NE12	NE13	NE14	NE15
	NE16	NE17	NE18	NE19	NE20
Continue Auth Session	0x00				
Authorization Digest	AD1	AD2	AD3	AD4	AD5
	AD6	AD7	AD8	AD9	AD10
	AD11	AD12	AD13	AD14	AD15
	AD16	AD17	AD18	AD19	AD20

Figure 14.24 TPM_AuthorizeMigrationKey output message block

In conclusion, the TPM Owner must be involved during the creation of a migration blob. To avoid explicit Owner authorization within the TPM_CreateMigrationBlob, a TCG_MIGRATIONKEYAUTH structure is established by the TPM Owner and referenced during migration blob creation. The TCG_MIGRATIONKEYAUTH structure uses the TPM Proof, a unique secret inside every TPM, to force the use of this structure and associated authorization to a specific TPM – the TPM that created the structure. For more information regarding Owner commands not discussed in this chapter, refer to the TPG Main Specification, Version 1.1b.

The Key Management Command Suite

This chapter concerns the Trusted Platform Module (TPM) commands that affect the internal key management from the perspective of the types of keying material generated internally when these keys are loaded, getting the public key of keying material stored within the TPM, and removing keys from the TPM. There are basically two types of keying material that can be stored within the TPM: externally generated keys or internally generated keys. Did you say internally generated keys? If the keys are generated internally, why do they have to be loaded?

This can be a very confusing issue and the concept is found within the two commands that provide the ability to create and load the TPM internally generated keys: the TPM_CreateWrapKey and TPM_LoadKey. The idea is one would use the TPM facilities to internally generate and wrap the key, protect the private data using a parent key, prior to returning the generated key back to the user. The user can now do what she wishes with this keying material – turn right around and load this key into the TPM, the wrapped key is in a format compliant with the TPM_LoadKey command input block specifications, or save this data off for later use. This gives flexibility regarding when a key is generated internally and when that internally generated key is loaded into the TPM. This is no different than having a system whose sole purpose is to generate RSA key pairs. The system doesn't know where the keys will be loaded, just that other systems can use these keys when they choose to do so. This system can leverage the services of a single TPM to generate keying material for numerous TPM-based systems that are physically located externally from the key generation system.

15.1 The TPM_CreateWrapKey Command

Let's start off the discussion by considering the command TPM_CreateWrapKey. The command input message block is defined in Figure 15.1. This command, like all TPM commands, has the standard header information comprised of the authorization tag, this is a single authorized command, parameter size, and command ordinal.

The next parameter is defined as TCG_KEY_HANDLE, which points to a TPM internally stored RSA key of type TPM_KEY_STORE. If this Key Handle does not index this type of RSA key, the command will return the error TCG_INVALID_KEYUSE. The reason for this is that the Key Handle points to a parent key with regard to the key about to be created internally, and the definition of a TPM_KEY_STORE type is one that is used to wrap keying material – hence, a parent key.

The next two parameters are about the key that is being created, Usage Secret and Migration Secret – two secrets used to authorize the keys usage and migration. These values cannot be sent in clear, unless you want everyone and your brother to know what these values are (not a good idea). The solution is to encrypt these values leveraging the **EncAuth**; its use is defined in Chapter 10 about atypical command authorization. This is simply an eXclusive OR (XOR) of the data to protect with a "generated mask" that is the same size as the data to encrypt.

After this is done, the last step, with regard to the command input payload area, is to define what type of key you would like to create using the TPM_KEY structure. The TCG_KEY structure is defined in Figure 15.2; this structure is described in a rigorous fashion, including Platform Configuration Register (PCR) inclusion, since this command is responsible for key generation and could very well be a gating issue concerning every TPM-based command that involves RSA key use.

The first variable, regarding this Trusted Computing Group (TCG) structure type, involves the TCG_VERSION, which is the version of the TCG specification and the TPM firmware revision. To obtain the value, one must perform a TPM_GetCapability command with the following **CapArea** TCG_CAP_VERSION, as defined in Chapter 20 about this command; the TCG_VERSION type is defined again in Figure 15.3.

The next parameter involves the TCG_KEY_USAGE and, regarding the TPM_CreateWrapKey, will signify what subtype of key is to be created as

Authorization Tag	0x00	0xC2			
Parameter Size	0x00	0x00	0x00	0x92	
Ordinal	0x00	0x00	0x00	0x1F	
Key Handle	0x00	0x00	0x00	0x00	
EncAuth Usage Secret	EU1	EU2	EU3	EU4	EU5
	EU6	EU7	EU8	EU9	EU10
	EU11	EU12	EU13	EU14	EU15
	EU16	EU17	EU18	EU19	EU20
EncAuth Migration Secret	MU1	MU2	MU3	MU4	MU5
	MU6	MU7	MU8	MU9	MU10
	MU11	MU12	MU13	MU14	MU15
	MU16	MU17	MU18	MU19	MU20
TCG_KEY, Key to Create	0x01	0x01	0x00	0x06	
	0x00	0x10			
	0x00	0x00	0x00	0x02	
	0x01				
	0x00	0x00	0x00	0x01	
	0x00	0x01			
	0x00	0x02			
	0x00	0x00	0x00	0x0C	
	0x00	0x00	0x80	0x00	
	0x00	0x00	0x00	0x02	
	0x00	0x00	0x00	0x00	
	0x00	0x00	0x00	0x18	
Digest at Release	DR1	DR2	DR3	DR4	DR5
	DR6	DR7	DR8	DR9	DR10
	DR11	DR12	DR13	DR14	DR15
	DR16	DR17	DR18	DR19	DR20
Digest at Creation	DC1	DC2	DC3	DC4	DC5
	DC6	DC7	DC8	DC9	DC10
	DC11	DC12	DC13	DC14	DC15
	DC16	DC17	DC18	DC19	DC20
	0x00	0x00	0x00	0x00	
Authorization Handle	0x00	0x00	0x00	0X00	
nonceOdd	NO1	NO2	NO3	NO4	NO5
	NO6	NO7	NO8	NO9	NO10
	NO11	NO12	NO13	NO14	NO15
	NO16	NO17	NO18	NO19	NO20
Continue Auth Session	0x00				
Authorization Digest	AD1	AD2	AD3	AD4	AD5
	AD6	AD7	AD8	AD9	AD10
	AD11	AD12	AD13	AD14	AD15
	AD16	AD17	AD18	AD19	AD20

Figure 15.1 TPM_CreateWrapKey input message block

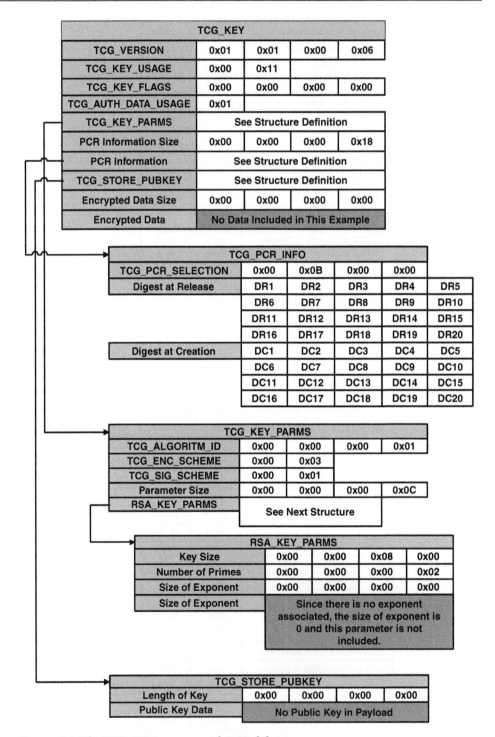

Figure 15.2 The TCG_KEY structure with PCR definition

TCG_VERSION	
Major	Version Byte, Usually the TCG Spec
Minor	Version Byte, Usually the TCG Spec
Revision Major	Byte Representing the FW Version
Revision Minor	Byte Representing the FW Version

Figure 15.3   The TCG_VERSION type, TPM_CreateWrapKey

TCG_KEY_USAGE	
TPM_KEY_SIGNING	Signing Key
TPM_KEY_STORAGE	Storage Key
TPM_KEY_IDENTITY	Identity Key
TPM_KEY_AUTHCHANGE	Authorization Change Key
TPM_KEY_BIND	Binding Key
TPM_KEY_LEGACY	Legacy Key

Figure 15.4   The TCG_KEY_USAGE type, TPM_CreateWrapKey

the result of this command's execution. Note the base type of the key to be created is always an RSA key. See Figure 15.4 concerning the types of key usage values that are legal to use concerning the TPM_CreateWrapKey command execution.

Notice that there are two key types that will cause the TPM_Create-WrapKey to fail with regard to command execution: **TPM_KEY_AUTHCHANGE** and **TPM_KEY_IDENTITY**. These types are considered "specialty" types and must be used within the command context that specifically operates with keys of this type. In addition, key generation regarding **TCG_KEY_USAGE** will limit the RSA operation(s) any one key can perform regarding the specific. For example, an RSA key generated by the TPM_CreateWrapKey execution with a **TCG_KEY_USAGE** of **TPM_KEY_SIGNING** will only be able to perform cryptographic functions related to an RSA sign. This particular key, when eventually loaded within the TPM, will not be able to, for example, wrap private data associated with a child key – the function of a **TPM_KEY_STORAGE** key. The only exception to this rule would concern the **TPM_KEY_LEGACY** type, allows RSA encrypt and signature operation to be performed but is considered a deprecated key type and must be used as such – recommended.

The next parameter within the **TCG_KEY** structure concerns the authorization, TPM_TakeOwnership command. Figure 15.5 describes this parameter option, which involves two choices: authorized use or non-authorized use. This option has a little more meaning, outside of the definition regarding the Storage Root Key (SRK), and will allow the keying material that is about to be generated to require authorization prior to its use or not.

TCG_AUTH_DATA_USAGE	
Authorization Not Required	0x00
Authorization Required	0x01

Figure 15.5 The TCG_AUTH_DATA_USAGE type, TPM_CreateWrapKey

The **TCG_KEY_PARMS**, the next parameter to be considered, concerns the RSA specific key generation inputs. In other words, when the keying material is generated, as a result of TPM_CreateWrapKey command execution, the TPM must understand the type of keying material to be generated. In the case of this command, the only type of key that can be created is RSA, defined by the **TCG_ALGORITHM_ID**. Figure 15.6 defines this structure.

Here we come to some choices regarding the key that is about to be generated. You must know the size and flavor, signing, storage, or other type of key defined by **TCG_KEY_USAGE**. In addition, the encryption and signature schemes must be defined regarding key usage. Note that both schemes must be defined to a valid scheme identifier or none, indicating that the **TCG_KEY_USAGE** type does not need this scheme defined. Figure 15.7 defines the possible schemes regarding both encryption and signature.

The last parameter, regarding the **TCG_KEY_PARMS**, concerns the algorithm specific structure **RSA_KEY_PARMS**; in the case of the Atmel TPM, the

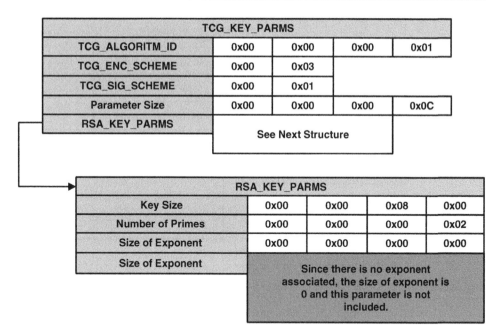

Figure 15.6 The TCG_KEY_PARMS structure, TPM_CreateWrapKey

TCG Encryption Schemes	
TCG_ES_NONE	0x0001
TCG_ES_RSAESOKCSv15	0x0002
TCG_ES_RSAESOAEP_SHA1_MGF 1	0x0003

TCG Signature Schemes	
TCG_SS_NONE	0x0001
TCG_SS_RSASSAPKCS1v15_SHA1	0x0002
TCG_SS_RSASSAPKCS1v15_DER	0x0003

Figure 15.7 The TCG_ENC_SCHEME and TCG_SIG_SCHEME, TPM_CreateWrapKey

TCG_KEY_PARMS will always pertain to RSA. Figure 15.8 defines the parameters contained within this type of structure.

There are no real surprises here; the structure will define the RSA key size, number of primes, and exponent related parameters. A few notes: the

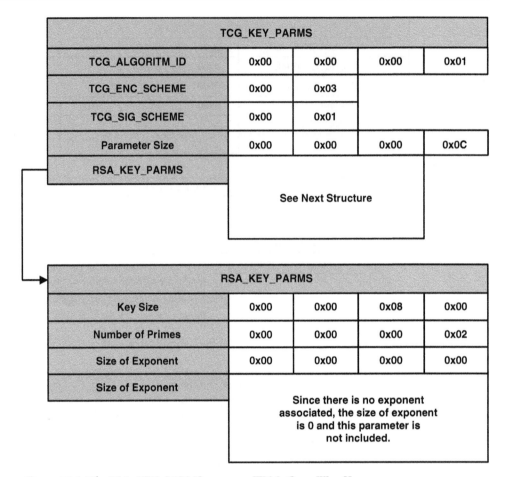

Figure 15.8 The RSA_KEY_PARMS structure, TPM_CreateWrapKey

number of primes will always be two and the exponent data will be set to *None* by making the exponent size zero. Figure 15.9 shows some example **TCG_KEY_PARMS** structures, which define various types to RSA keys that will be created for the execution of TPM_CreateWrapKey.

0x00	0x00	0x00	0x01
0x00	0x01		
0x00	0x02		
0x00	0x00	0x00	0x0C
0x00	0x00	0x80	0x00
0x00	0x00	0x00	0x02
0x00	0x00	0x00	0x00

Figure 15.9 TCG_KEY_PARMS structure data population example

Here we come to PCR-specific parameters; these parameters encompass the **PCRInfoSize** along with the **PCRInfo** which determines whether the key about to be generated is tied to PCR values and if so, which ones. The TPM_ TakeOwnership command chapter describes the specific procedure for defining the key generation regarding no PCR value attachment. Simply, the parameter **PCRInfoSize** is set to a value of zero, which indicates that there is no PCR value attached to this key. We are not so lucky in this chapter's description and the key about to be generated will have PCR values attached to the key being generated. Figure 15.10 poses a description concerning PCR attachment.

	0x00	0x00	0x00	0x18	
Digest at Release	DR1	DR2	DR3	DR4	DR5
	DR6	DR7	DR8	DR9	DR10
	DR11	DR12	DR13	DR14	DR15
	DR16	DR17	DR18	DR19	DR20
Digest at Creation	DC1	DC2	DC3	DC4	DC5
	DC6	DC7	DC8	DC9	DC10
	DC11	DC12	DC13	DC14	DC15
	DC16	DC17	DC18	DC19	DC20

Figure 15.10 The PCRInfoSize and PCRInfo parameter data population

Notice that the **PCRInfoSize** does not contain a value of zero, but defines the size of the resulting **PCRInfo**, which is a structure of type TCG_PCR_ INFO. Figure 15.11 gives an example of the **PCRInfo** structure that relates to the PCR assignment regarding PCRs zero, one, and three.

TCG_PCR_INFO					
TCG_PCR_SELECTION	0x00	0x0B	0x00	0x00	
Digest at Release	DR1	DR2	DR3	DR4	DR5
	DR6	DR7	DR8	DR9	DR10
	DR11	DR12	DR13	DR14	DR15
	DR16	DR17	DR18	DR19	DR20
Digest at Creation	DC1	DC2	DC3	DC4	DC5
	DC6	DC7	DC8	DC9	DC10
	DC11	DC12	DC13	DC14	DC15
	DC16	DC17	DC18	DC19	DC20

Figure 15.11 A PCRInfo structure binding PCRs 0, 1, and 3

The first parameter is that of a TCG_PCR_SELECTION structure type and defines the size of the PCR selection bit map, 2 bytes indicate 16 PCR indices and are defined by **size of Select**. The last parameter involves the selection of the PCRs to be bound and this information is contained within the bytes, 16 bits indicating the inclusion of each of the possible 16 PCR values. The format of each of the **pcrSelect** bits concerns the logic of each bit; a False (0) indicates that the indexed PCR will not be included and a True (1) indicates that the PCR will be included. The resulting TCG_COMPOSITE_HASH values will be stored within two TCG_PCR_INFO structure parameters: the **digestAtRelease** and the **digestAtCreation**. Figure 15.12 defines the calculation algorithm regarding the TCG_COMPOSITE_HASH digest calculation.

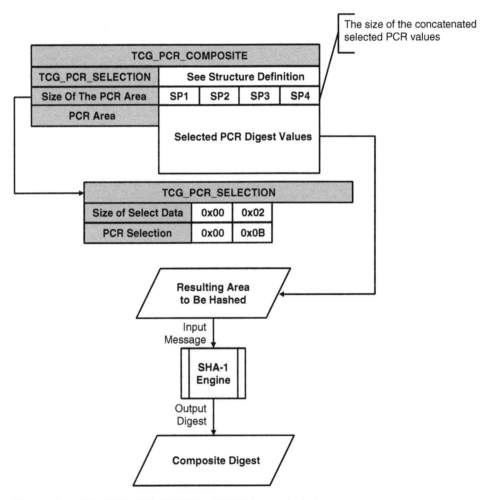

Figure 15.12 The TCG_COMPOSITE_HASH digest calculation

When the PCR-specific data is compiled, it is inserted into the TCG_KEY structure and the key being generated, in lieu of the TPM_CreateWrapKey execution, will be tied to the digests stored in PCRs zero, one, and three. Consequently, if the key being created is loaded into the TPM and PCR zero, one, or three has been modified since the time of key creation, the TPM_LoadKey command will fail with a command return code TPM_INVALID_PCR_INFO.

The last remaining TCG_KEY structure parameters are the RSA public key and private encrypted data, which will be set to zeros, no data included, since the information described by this data has not been defined. This data has not been defined because the purpose of the TPM_CreateWrapKey is to generate an RSA key. Figure 15.13 describes the two remaining parameter structures concerning TCG_STORE_PUBKEY: **encSize** and **encData**. Remember that there is no **encData** associated at this time and therefore, the only data parameter needed to convey this parameter information is **encSize**, which is defined as zero or 0x00000000.

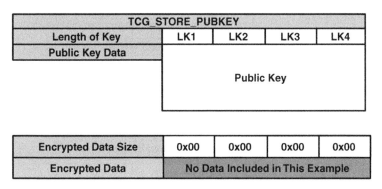

Figure 15.13 The TCG_STORE_PUBKEY, encSize, and encData parameters

The final task, with regard to the TPM_CreateWrapKey input message is to attach the authorization block. Here comes the interesting issue concerning this command; the authorization session, indexed by the authorization handle contained within the authorization block, must be an Object Specific Authorization Protocol (OSAP). The reason for this, regarding TPM_Create-WrapKey command execution, is that the **EncAuth** used to encrypt the private data held within the command input message depends on a Shared Secret. Remember the only authorization protocol that issues a Shared Secret is the OSAP and if an Object Independent Authorization Protocol (OIAP) authorization session is referenced, the command will not be able to calculate the **EncAuth** value(s). If the authorization session is indeed an OIAP, oops, the command execution will fail, returning a response code of **TPM_AUTHFAIL**. The moral of the story: do not attempt to execute the TPM_CreateWrapKey command referencing an OIAP.

Now that we have compiled the command input message block and the TPM has performed the CreateWrapKey functionality, the execution results are available for us to recover. The output command message block contains the usual authorization tag, parameter size, and return code. The only command payload data parameter is the internally generated key as described within the **TCG_KEY** type structure. The entire TPM_CreateWrapKey output command message block is defined in Figure 15.14.

Authorization Tag	0x00	0xC5			
Parameter Size	PS1	PS2	PS3	PS4	
Return Code	0x00	0x00	0x00	0x00	
TCG_PUBKEY	0x00	0x00	0x00	0x01	
	0x00	0x01			
	0x00	0x02			
	0x00	0x00	0x00	0x0C	
	0x00	0x00	0x80	0x00	
	0x00	0x00	0x00	0x02	
	0x00	0x00	0x00	0x00	
	Public Key That Was Created and Wrapped				
Authorization Handle	0x00	0x00	0x00	0X00	
nonceEven	NE1	NE2	NE3	NE4	NE5
	NE6	NE7	NE8	NE9	NE10
	NE11	NE12	NE13	NE14	NE15
	NE16	NE17	NE18	NE19	NE20
Continue Auth Session	0x00				
Authorization Digest	AD1	AD2	AD3	AD4	AD5
	AD6	AD7	AD8	AD9	AD10
	AD11	AD12	AD13	AD14	AD15
	AD16	AD17	AD18	AD19	AD20

Figure 15.14 TPM_CreateWrapKey output message block

Note that the command output payload, concerning the TPM_ CreateWrapKey, can be used directly by the TPM_LoadKey with regard to its input command message block. Notice that the only input parameter defined in the TPM_LoadKey input message block is that of a **TCG_KEY** type structure. Refer to the section about the command TPM_LoadKey for more information. The output concerning TPM_CreateWrapKey is identical, regarding payload, to that of the TPM_TakeOwnership command. The only exception concerns the resulting keying material, regarding TPM_CreateWrapKey, is tied to PCRs, and the TPM_TakeOwnership SRK contained within its output message payload is not. By juxtaposing the output message block, specifically the payload, with regard to the TPM_CreateWrapKey and TPM_ TakeOwnership, one can see these differences. The point is that the output payload is of type **TCG_KEY** and the specific parameter definition can be found in the chapter about TPM_TakeOwnership.

Finally, this command output is single authorized and an authorization block is attached. This authorization block references an OSAP authorization session and the TCG Software Stack (TSS) must use the Shared Secret to authenticate the command output message block.

15.2   The TPM_LoadKey Command

The TPM_LoadKey is a command whose sole purpose is to get keying material into the TPM of the RSA type. With that said, let's look at the command input message block, as defined in Figure 15.15. This command input block has the typical authorization tag, parameter size, and command ordinal. The command payload has two input parameters: the Parent Handle and the Key to be loaded in the form of **TCG_KEY** structure type.

This structure has been "beaten-to-death" and if you need to review its details, see the TPM_TakeOwnership or TPM_CreateWrapKey command discussion. This command has a single authorization block and can be of type OIAP or OSAP, with the authorization digest Hash-based Message Authentication Code (HMAC) key being the Usage Secret of the parent key or the Shared Secret, respectively.

The command output message block is defined in Figure 15.16. Same beginning header information, regarding the output command message, concerns the authorization tag, parameter size, and the command return code. The only output payload parameter, regarding the TPM_LoadKey command,

Authorization Tag	2 bytes	0x00	0xC2		
Parameter Size	4 bytes	0x00	0x00	0x02	0x6a
Ordinal	4 bytes	0x00	0x00	0x00	0x20
Key Handle	4 bytes	0x40	0x00	0x00	0x00
TCG Key Structure	555 bytes	0x01	0x01	0x00	0x00
		0x00	0x10		
		0x00	0x00	0x00	0x02
		0x01			
		0x00	0x00	0x00	0x01
		0x00	0x01		
		0x00	0x03		
		0x00	0x00	0x00	0x0c
		0x00	0x00	0x80	0x00
		0x00	0x00	0x00	0x02
		0x00	0x00	0x80	0x00
		0x00	0x00	0x80	0x00
		0x00	0x00	0x01	0x00
		Encrypted Blob Reference Key Handle			
		0x00	0x00	0x01	0x00
		Encrypted Blob Reference Key Handle			
Authorization Handle	4 bytes	0x00	0x00	0x00	0x00
nonceOdd	20 bytes	NO	NO	NO	NO	NO
		NO	NO	NO	NO	NO
		NO	NO	NO	NO	NO
		NO	NO	NO	NO	NO
Continue Auth Session	1 byte	0x00			
Authorization Digest	20 bytes	AD	AD	AD	AD	AD
		AD	AD	AD	AD	AD
		AD	AD	AD	AD	AD
		AD	AD	AD	AD	AD

Figure 15.15 TPM_LoadKey input message block

has to do with a TCG_KEY_HANDLE, which, on successful command execution, references the keying material loaded in the TPM's internal boundary. The size of the TCG_KEY_HANDLE is 4 bytes.

The output message block has an authorization block attached, which is relative to the entity whose use is being authorized – the parent key referenced by the TCG_KEY_HANDLE. In the case that the parent key usage was defined such that it didn't need authorization, this command would be

Authorization Tag	2 bytes	0x00	0xC5			
Parameter Size	4 bytes	PS1	PS2	PS3	PS4	
Return Code	4 bytes	0x00	0x00	0x00	0x00	
Key Handle	4 bytes	0x00	0x00	0x00	0x01	
nonceEven	20 bytes	NE	NE	NE	NE	NE
		NE	NE	NE	NE	NE
		NE	NE	NE	NE	NE
		NE	NE	NE	NE	NE
Continue Auth Session	1 byte	0x00				
Authorization Digest	20 bytes	AD	AD	AD	AD	AD
		AD	AD	AD	AD	AD
		AD	AD	AD	AD	AD
		AD	AD	AD	AD	AD

Figure 15.16 TPM_LoadKey output message block

unauthorized. If you forget this case and try to authorize a TPM_LoadKey command in which the entity being referenced, in this case the parent key that was set to require no authorization, the command will fail with a command code of **TPM_AUTHFAIL**. You must abide by the authorization attributes assigned to each entity held within the TPM regardless of authorization level.

Some of the failure response codes you might come across when trying to execute this command are discussed in the following paragraph. First, and obviously, the parent key used to wrap the child key about to be loaded into the TPM must be resident within the TPM internal boundaries. If not, the TPM will respond with an error return code stating that the TPM contains no such key, **TCG_KEYNOTFOUND**. Considering that the only type of key that can be loaded in the Atmel TPM is of type RSA, the TPM will error with **TCG_BAD_KEY_PROPERTY** if the key about to be loaded in not of the type RSA.

We are trying to load a child key into TPM; therefore the type of key the parent key must be concerned with is a **TPM_KEY_STORAGE**. If the parent key is some other type of **TCG_KEY_USAGE**, the TPM will fail indicating a **TPM_INVALID_KEY_USAGE** error. The key about to be loaded within the TPM must not be of the **TCG_KEY_USAGE** type concerning **TPM_KEY_IDENTITY** or **TPM_KEY_AUTHCHANGE**; if so, the TPM will fail with a return code of **TPM_INVALID_KEY_USAGE**. These key usage types are used within the context of specific TPM functionality and cannot be generically loaded into the TPM outside of the specific command execution boundaries.

The TPM has a finite number of keys that can be loaded; this depends on the TPM memory size, key sizes, and number of keys existing in the TPM at the time of trying to load another key. If the TPM has no room to store the key that you are attempting to load, the TPM will respond with the error, TCG_NOSPACE. This error is recoverable, but this involves the deletion of existing keying material, internal to the TPM, prior to resending the TPM_LoadKey command.

With that said, let's look at the command that will accomplish the goal of removing keying material from the TPM, which has been loaded at various TPM_LoadKey execution events.

15.3   The TPM_EvictKey Command

This command execution is very straightforward; get the key, referenced by the TCG_KEY_HANDLE, out of the TPM internal storage. The TPM_EvictKey input command message block is defined in Figure 15.17.

This command has a very simple parameter list. The usual header information in regard to the authorization tag, no authorization required, parameters size, and command ordinal. The only input command payload parameter concerns a TCG_KEY_HANDLE that references the key about to be deleted from the TPM internal memory.

The resulting command output message block, after successful command execution, is shown in Figure 15.18. This output response is very simple; the command either passed or had some type of error. One of the most relevant command errors involves the reference to a key that is not stored in the TPM. If this is the case, the TPM will error with a return result of TCG_INVALID_KEYHANDLE. That's it for this command.

Authorization Tag	2 bytes	0x00	0xC1		
Parameter Size	4 bytes	0x00	0x00	0x00	0x0E
Ordinal	4 bytes	0x00	0x00	0x00	0x22
Key Handle	4 bytes	0x00	0x00	0x00	0x01

Figure 15.17 TPM_EvictKey input message block

Authorization Tag	2 bytes	0x00	0xC4		
Parameter Size	4 bytes	0x00	0x00	0x00	0x0A
Ordinal	4 bytes	0x00	0x00	0x00	0x21

Figure 15.18 TPM_EvictKey output message block

The last command to be considered in this chapter involves the public key associated with the key set stored in the TPM and obtaining its information.

15.4 The TPM_GetPubKey Command

Once again, a very simple command presents a `TCG_KEY_HANDLE` on input and returns the public portion of a TPM internally stored RSA key using the `TCG_PUBKEY` structure type format. The input command message block concerning the TPM_GetPubKey is defined in Figure 15.19.

Authorization Tag	2 bytes	0x00	0xC2			
Parameter Size	4 bytes	PS1	PS2	PS3	PS4	
Ordinal	4 bytes	0x00	0x00	0x00	0x21	
Key Handle	4 bytes	0x00	0x00	0x00	0x01	
nonceOdd	20 bytes	NO	NO	NO	NO	NO
		NO	NO	NO	NO	NO
		NO	NO	NO	NO	NO
		NO	NO	NO	NO	NO
Continue Auth Session	1 byte	0x00				
Authorization Digest	20 bytes	AD	AD	AD	AD	AD
		AD	AD	AD	AD	AD
		AD	AD	AD	AD	AD
		AD	AD	AD	AD	AD

Figure 15.19 TPM_GetPubKey input message block

This command has the usual culprits with the additional payload parameter defined as a `TCG_KEY_HANDLE` type. This parameter references a key stored internally within the TPM. This command is a single authorization command; we must authorize the entity referenced by the `TCG_KEY_HANDLE` prior to getting the public key, regarding the referenced key, on command output.

The successful command execution will result with the output message block defined in Figure 15.20. Same old, same old output message block header information; this is a single authorization command with an additional payload parameter. The only payload parameter, regarding TPM_GetPubKey, concerns the public portion of the RSA key referenced by the `TCG_KEY_HANDLE` and contained in the `TCG_PUBKEY` type structure.

Authorization Tag	2 bytes	0x00	0xC2			
Parameter Size	4 bytes	PS1	PS2	PS3	PS4	
Ordinal	4 bytes	0x00	0x00	0x00	0x21	
Key Handle	4 bytes	0x00	0x00	0x00	0x01	
nonceOdd	20 bytes	NO	NO	NO	NO	NO
		NO	NO	NO	NO	NO
		NO	NO	NO	NO	NO
		NO	NO	NO	NO	NO
Continue Auth Session	1 byte	0x00				
Authorization Digest	20 bytes	AD	AD	AD	AD	AD
		AD	AD	AD	AD	AD
		AD	AD	AD	AD	AD
		AD	AD	AD	AD	AD

Figure 15.20 TPM_GetPubKey output message block

Figure 15.21 defines the **TCG_PUBKEY** structure type, which contains two parameter definitions: the **TCG_KEY_PARMS** and the **TCG_STORE_PUBKEY**; both of these are defined in Figure 15.22.

The **TCG_KEY_PARMS** have been referenced numerous times throughout this book, and the key information regarding the algorithm ID has been defined: encryption scheme, signature scheme, key length, number of primes, and exponent information attached to the public key that was returned. The **TCG_STORE_PUBKEY** has also been discussed and defines the actual public key data associated with the **TCG_KEY_HANDLE** and in the form of key length and key.

In summary, the TPM has the means to internally generate RSA key pairs and return those keys to the entity making this function call, the TPM_ CreateWrapKey. Once these key(s) are created, they can be loaded within the TPM internal memory in regard to the command TPM_LoadKey. If TPM internal space limitations are breached, some or all of the keying material, which was loaded into the TPM, can be deleted using the command TPM_EvictKey. The public portion of any RSA key, which resides within the TPM, excluding the Endorsement Key (EK), can be requested by the use of the command TPM_GetPubKey.

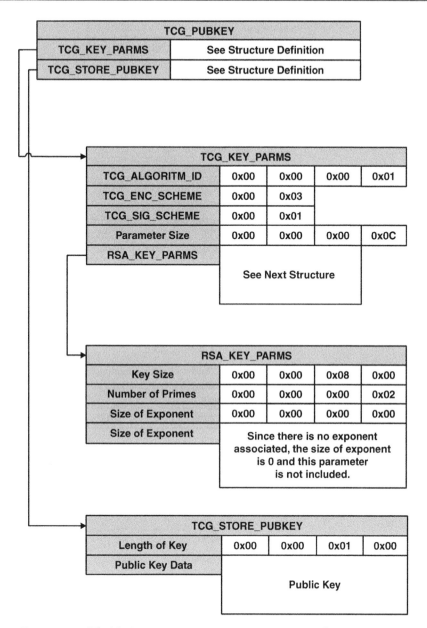

Figure 15.21 The TCG_PUBKEY structure type, TPM_GetPubKey

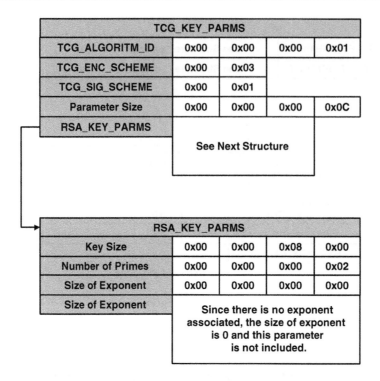

TCG_KEY_PARMS				
TCG_ALGORITM_ID	0x00	0x00	0x00	0x01
TCG_ENC_SCHEME	0x00	0x03		
TCG_SIG_SCHEME	0x00	0x01		
Parameter Size	0x00	0x00	0x00	0x0C
RSA_KEY_PARMS	See Next Structure			

RSA_KEY_PARMS				
Key Size	0x00	0x00	0x08	0x00
Number of Primes	0x00	0x00	0x00	0x02
Size of Exponent	0x00	0x00	0x00	0x00
Size of Exponent	Since there is no exponent associated, the size of exponent is 0 and this parameter is not included.			

TCG_STORE_PUBKEY				
Length of Key	0x00	0x00	0x01	0x00
Public Key Data	Public Key			

Figure 15.22 The TCG_KEY_PARMS and TCG_STORE_PUBKEY, TPM_GetPubKey

16

The RSA Encryption and Decryption Command Suite

This chapter concerns the Trusted Platform Module (TPM) Command Suite that deals with the encryption and decryption of data using the RSA key(s) stored within the TPM device. The first command is TPM_UnBind, which is a flavor of RSA decryption. A side note: the only TPM command supported, when considering "binding" operations, is the TPM_UnBind command. There exists bind functionality, but this TSS_Bind command is defined within the TCG Software Stack (TSS) hence the command name. In addition, keying material stored within the TPM and used to bind data to that key may be migrated from one TPM, A, to another TPM, B, and the command TPM_UnBind executed using TPM B to get at the data. The point is that a **TPM_KEY_BIND** key type, used in conjunction with the command TPM_UnBind, can be migrated to another TPM if the **TCG_KEY_FLAGS** associated with this key indicate that the key is migratable.

The other type of encryption supported by the TPM uses the two commands: TPM_Seal and TPM_UnSeal. These commands have a slightly different meaning than the TSS_Bind and TPM_UnBind commands, which encrypt data relative to keying material stored within the TPM. The two flavors of Seal commands also encrypt data, but cryptographically "seal" or lock this data within a single TPM device; hence the term *seal*. This would be no different than if you had the ability to physically store unlimited data within a TPM device. The data would exist only within the context of that individual TPM and only that one; therefore the keying material associated with the Seal commands must be nonmigratable.

In addition, within TPM_Seal and TPM_UnSeal, there are two authorization levels that must be satisfied: the Usage Secret concerning the RSA encryption key and Authorization Secret is associated with the data itself. This implies that when sealing a message, the command issuer must know the Usage Secret for the RSA encryption key and supply an Authorization Secret that can be associated with the encrypted blob that is generated. Therefore, when the command TPM_UnSeal is issued relative to the encrypted blob produced by TPM_Seal, two authorization sessions must be implemented. One session to authorize the use of the RSA decryption key and another to authorize the use of the data that will be returned on successful command execution.

Therefore we have four commands that handle the encryption and decryption of data coming into and going out of the TPM. Three commands are supported by the TPM: TPM_UnBind, TPM_Seal, and TPM_UnSeal. One command is supported by the TSS, TSS_Bind. All four command descriptions are covered in this chapter, including the TSS_Bind command. The TSS_Bind is worth investigating since some readers might want to know about a Secure Software Stack that can bind data to keying material stored within the TPM they are communicating with. With that said, let's look at the TSS_Bind command definition first.

16.1 The TSS_Bind or Tspi_Data_Bind (TSS Specification)

This command is very different with regard to description and Application Programming Interface (API) because it is defined within the TSS specification. The interface is designed, using a TSS API call, to produce the TSS_Bind functionality that will create an encrypted blob that is cryptographically associated with a corresponding private key stored within the TPM. This does not mean that readers must design a fully functioning TSS to use this function, but they may wish to add this functional support to their embedded application design. With that said, let's examine the API defined in the TSS Specification with regard to the TSS_Bind command, as shown in Figure 16.1.

This function definition is very straightforward; encrypt a message pointed to by **rgbDataToBind** whose length is **ulDataLength** with the RSA public key pointed to by **hEncKey** and store the encrypted blob to a memory location pointed to by **hEncData**.

Now let's discuss what this function does, with regard to the data that is to be encrypted, which is dependent on which Trusted Computing Group

TSS API Tspi_Data_Bind Function

```
TSS_RESULT Tspi_Data_Bind
(
    TSS_HENCDATA    hEncData,        // Handle pointing to resulting data blob.
    TSS_HKEY        hEncKey,         // Handle to RSA public key.
    UINT32          ulDataLength,    // Size of message to encrypt.
    BYTE*           rgbDataToBind    // Message to encrypt.
);
```

Figure 16.1 The TSS_Bind command API

(TCG) encryption scheme is defined for use: **TCG_ES_RSAESPKCSv15** or **TCG_ES_RSAESPKCSOAEP_SHA1_MGF1**. The RSA key loaded in the TPM that will perform the TPM_UnBind function determines the encryption scheme that is to be used with the **TCG_KEY_PARMS** associated with this keying material. The first encryption scheme we consider is the **TCG_ES_RSAESPKCSOAEP_ SHA1_MGF1**. This scheme simply encrypts the message, as is, with no further padding, in the form of TCG structures, applied. Therefore, encrypt the message with the selected RSA public key to produce the encrypted data blob that is to be associated with the TPM_UnBind command. See Figure 16.2 for a depiction of this procedure.

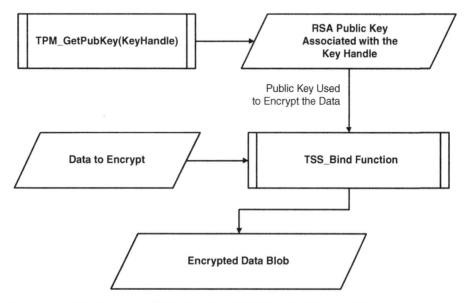

Figure 16.2 TSS_Bind using TCG_ES_RSAESPKCSOAEP_SHA1_MGF1 scheme

The result of the function, the encrypted blob, can be used directly within the TPM_UnBind command input message block. Make sure the TCG_KEY_PARMS associated with the RSA private key held within the TPM have the same attributes as the RSA public key used to encrypt the data.

The encryption scheme TCG_ES_RSAESPKCSv15 is different altogether concerning a TCG structure that the data to encrypt must be a parameter of introducing the TCG_BOUND_DATA structure type; this encrypted when binding data using the TCG_ES_RSAESPKCSv15 scheme. Figure 16.3 defines the structure parameters that must be populated to successfully bind the data using an RSA public key with the encryption scheme set to TCG_ES_RSAESPKCSv15.

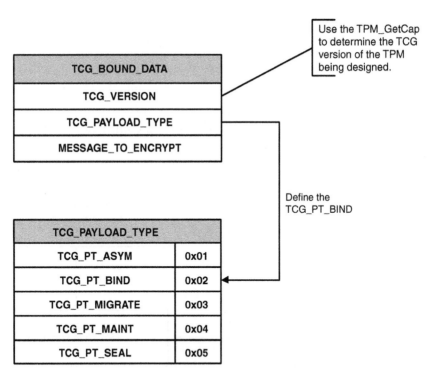

Figure 16.3   Binding with an encryption scheme of TCG_ES_RSAESPKCSv15

So, just like before, we encrypt the message using the RSA public key associated with the corresponding private key held within the TPM of type TPM_KEY_BIND. The only difference using the TCG_ES_RSAESPKCSv15

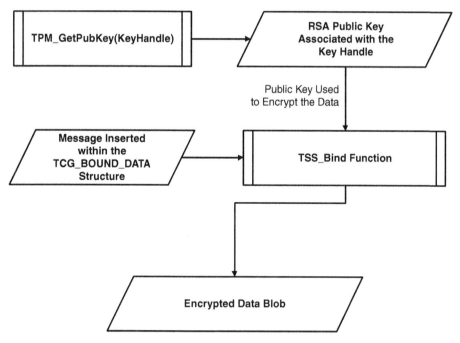

Figure 16.4 Binding with an encryption scheme of TCG_ES_RSAESPKCSOAEP_SHA1_MGF1

scheme is we are encrypting a TCG structure and not the raw message as we did using the **TCG_ES_RSAESPKCSOAEP_SHA1_MGF1**. See Figure 16.4, which defines this operation.

This scheme is similar concerning the RSA encryption using a public key associated with a corresponding private key held within the TPM. The only difference concerns the message itself; the **TCG_ES_RSAESPKCSOAEP_SHA1_MGF1** uses the raw message and the **TCG_ES_RSAESPKCSv15** uses the **TCG_BOUND_DATA** structure type to hold the message to encrypt.

That's it; this is the entire functional description concerning the TSS_Bind command defined by the TSS Specification. The sole purpose of this command is to create encrypted data blobs that are bound to private keys held within the TPM. The bind rules apply to the type of encryption scheme associated with the RSA key, specifically the private key held within the TPM and determined during the execution of the TPM_LoadKey command. Any mismatch with regard to the public to private key association or the encryption scheme requirements will result in the failure of the TPM_UnBind command. Now that we have a definition for the TSS_Bind functionality, let's look at the TPM_UnBind command.

16.2 The TPM_UnBind Command

This TPM command is used to unbind data that has been encrypted with an RSA public key whose corresponding private key is held within the TPM. The only legal types of keys that can be used by this command are a TPM_ KEY_BIND or a TPM_KEY_LEGACY, but we know how we all feel about Legacy Keys. If the Key Handle within the command input message block is pointing to a TPM_KEY_BIND key whose attributes include the encryption scheme TCG_ES_RSAESPKCSOAEP_SHA1_MGF1 or the key is defined as a TPM_KEY_ BIND, the decrypted message will be decrypted in and of itself. If the Key Handle is pointing to a TPM_KEY_BIND key whose attributes include the encryption scheme TCG_ES_RSAESPKCSv15, the decrypted message will be contained within the TCG_BOUND_DATA structure type. In this case, if this structure defines a TCG_PAYLOAD_TYPE other than TCG_PT_BIND, the TPM_ UnBind command will fail. This is reiterating the basic TSS_Bind encryption rules that are enforced during the execution of the TPM_UnBind command. If the TPM_UnBind command fails, check your TSS_Bind function for errors in the encryption procedure, including mismatched private and public keys, wrong encryption scheme, or general TPM_UnBind command input message errors. Figure 16.5 describes the TPM_UnBind command input message parameters.

There really isn't much to this command with regard to payload parameters; there is a Key Handle, encrypted data size, and the encrypted data itself. The usual command input message header information is included and this is indeed a single authorized command relative to the key entities Usage Secret. An authorization block is attached with regard to the authorization data that will authorize the use of the key referenced within the command input message and, as a result, allows the command execute. After successful execution, the TPM_UnBind command returns the output command message block as defined in Figure 16.6.

Again not much to discuss; the command output payload includes the return code, the decrypted data size, and the decrypted data along with the usual header information. The attached authorization block attests to the TSS that this is indeed a valid TPM command output response that allows the TSS to validate the message as such.

The idea behind the command TSS_Bind and TPM_UnBind is to allow data to be bound to an RSA private key existing within the TPM device. If

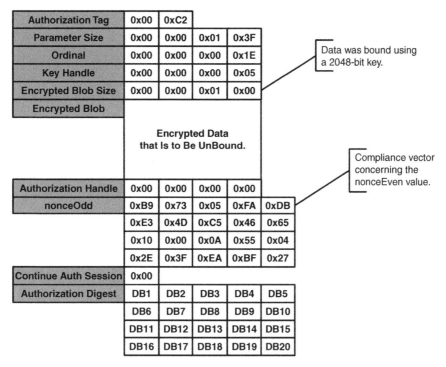

Figure 16.5 TPM_UnBind input message block

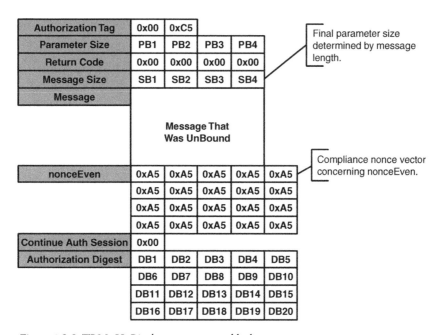

Figure 16.6 TPM_UnBind output message block

the RSA key attributes associated with the binding key allow migration (see the discussion concerning TPM_LoadKey), the binding key will be allowed to migrate to another TPM. This allows another TPM to perform a TPM_UnBind operation on the bound (encrypted) data since the binding property is applied to the RSA key not the TPM itself. Let's look at a situation where the encrypted data is bound to a unique TPM device itself and not just an RSA key. This concept is referred to "sealing" the data to a unique TPM, thus making it illegal to migrate the sealing key from one TPM device to another. With that said, the TPM_Seal command.

16.3   The TPM_Seal Command

The first observation that we can make, juxtaposing the TPM_Seal and TPM_UnSeal commands, are that the TPM_Seal command is a single authorized command and the TPM_UnSeal is a dual authorized command. We will get into the reasons why during the discussion concerning the TPM_UnSeal command. Let's look at the input message block for the TPM_Seal input command, shown in Figure 16.7.

The input command message block with regard to the TPM_Seal is more complicated for the TSS_Bind functionality concerning the TSS API. Not only does this command input identify the message size and information to be sealed but it also references **EncAuth** and Platform Configuration Register (PCR) information. The **EncAuth** is an encrypted 20-byte secret associated with the data to be encrypted, meaning not only do you have to authorize the unsealing of the data but also authorize the right to obtain the message itself. Hence, the dual authorizations concerning the TPM_UnSeal command. This command's input also references a Key Handle, which points to a sealing key (encryption key relative to RSA).

Regarding the PCR information, this means that the sealed data can be tied not only to a sealing key but also to the host system configuration. What this implies is that the data can be sealed relative to host system configuration via PCR composite digest, and on execution of the TPM_UnSeal command; if the host system's configuration has been altered, the TPM_UnSeal command will fail, returning a TCG_INVALID_PCR_INFO error. Remember, we are sealing this data to a unique TPM with dependencies defined by host system configuration.

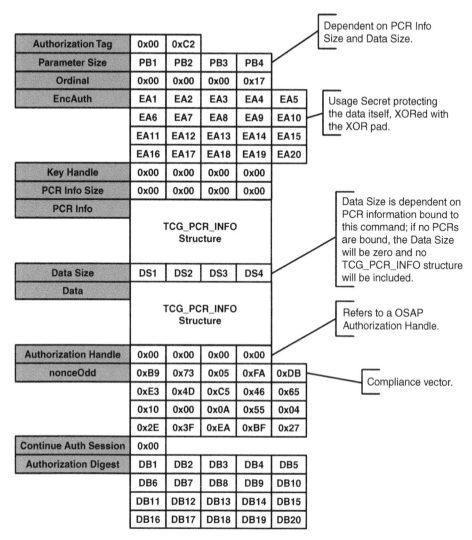

Figure 16.7 TPM_Seal input message block

The authorization session must be defined by the Object Specific Authorization Protocol (OSAP), which calculates a Shared Secret. This Shared Secret is used to produce an eXclusive OR (XOR) pad, which is used to encrypt the sealed data Authorization Secret. Calculation of the **EncAuth** is defined in Figure 16.8.

The **EncAuth** will be decrypted during TPM_Seal command execution and the data to be sealed will have an authorization protocol logically

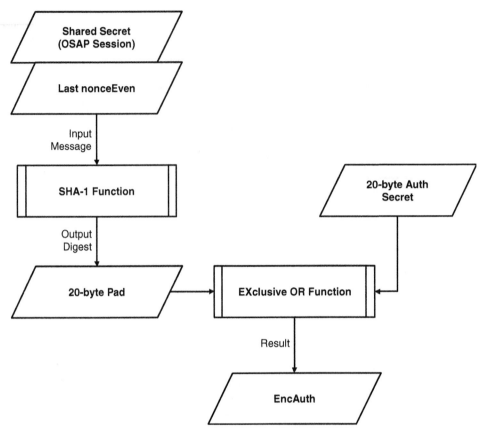

Figure 16.8 Calculating the EncAuth using the OSAP Shared Secret

wrapped around it prior to RSA encryption, based on the 20-byte decrypted secret protected by the **EncAuth**. The data protected by the 20-byte secret just described will then be encrypted using an internal RSA private key whose attributes are defined as **TPM_KEY_STORAGE** key type and is also defined as nonmigratable. Remember, we are sealing this data to a unique TPM; it wouldn't be so interesting if we could migrate the RSA key to another TPM.

The TPM_Seal output command message block is defined in Figure 16.9. The TPM_Seal command execution encrypts the data, which is additionally protected by an Authorization Secret using the RSA public key referenced by the Key Handle contained within the command input message block. If any PCR composite digests, determined by the **pcrInfo** (a PCR selection bit

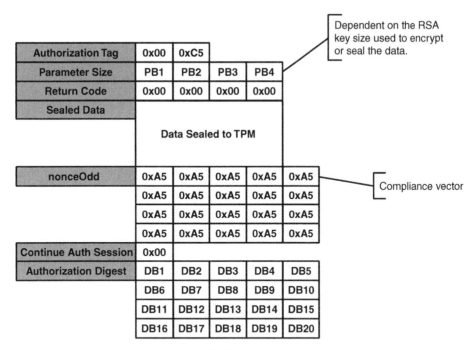

Figure 16.9 TPM_Seal output message block

field), are defined, the TPM will bind the data unsealing operation to the host system configuration measured by the composite digest. This command output message block has a single authorization block attached since the command requires the use of an authorized key entity.

The output payload contains a single parameter defined as a **TCG_STORED_DATA** structure type whose definition is shown in Figure 16.10. This **TCG_STORED_DATA** structure type is used within the TPM_UnSeal command input message block to identify the sealed data to be unsealed.

The TPM_Seal command locks an RSA public key encrypted data blob to a unique TPM, cryptographically storing the data in a particular TPM. The TPM_Seal command also creates a dependency with regard to host system configuration in the form of a composite digest, based on PCR selection. The TPM_Seal command produces a **TCG_STORED_DATA** structure type regarding the only payload parameter contained within the command output message block. This structure is included in the command input payload with regard to the TPM_UnSeal command.

TCG_STORED_DATA Structure Type	
TCG_VERSION	TCG Version contained within the TPM
Seal Info Size (UINT32)	Size of the Seal Info parameter (PCRs)
TCG_PCR_INFO	Structure type of TCG_PCR_INFO
Encrypted Data Size	Size of the encrypted data contained within the structure
TCG_SEALED_DATA	The encrypted data contained within the structure

Data Structure of This Type

TCG_SEALED_DATA Structure Type	
TCG_PAYLOAD_TYPE	Contains the type identifier TCG_PT_SEAL
TCG_SECRET	The decrypted authorization digest for the data
TCG_NONCE	An internal TPM Proof, unique 20-byte value
TCG_DIGEST	A digest of the TCG_STORED_DATA, excluding the parameters Encrypted Data Size and TCG_SEALED_DATA
Data Size (UINT32)	Size of the data to be sealed
Data	Data to be sealed

Figure 16.10 The TCG_STORED_DATA structure

16.4 The TPM_UnSeal Command

The TPM_UnSeal is the bookend to the TPM_Seal command and provides the capability regarding the decryption of the data sealed to the TPM. This command is a dual authorized command, since the decryption and data content must be authorized for use. Hence, the TPM_UnSeal is a form of deferred authorization, because the authorization digest, which is established during TPM_Seal command execution, needs to be determined before the TPM_UnSeal command can be fully authorized. Therefore the TPM authorization engine must authorize the use of the RSA decryption key and the "right-to-know" for the message that is about to be decrypted. All of the metrics that need to be known concerning the TPM_UnSeal command execution are conveyed within the encrypted TCG_STORED_DATA structure; see Figure 16.11 regarding the parameters contained within the command input message.

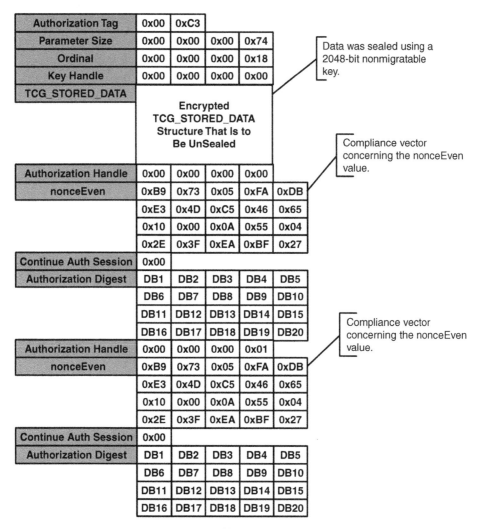

Figure 16.11 TPM_UnSeal input message block

One can see that the TPM_UnSeal is very much like the TPM_UnBind command output message block. The output payload contains the secret data size and the secret, which were sealed concerning the related TPM_Seal command. Notice the usual header information and that the dual authorization blocks relative to the encryption key Usage Secret and the secret determining the right-to-know this data's content. The TPM_UnSeal typically uses two Object Independent Authorization Protocol (OIAP) sessions, although one could use an OSAP regarding the authorization session tied to the decryption key.

Note that the authorization session associated with the message about to be decrypted must be an OIAP session. This is due to the fact that the authorization protocol used in determining the authenticity of the right-to-know involves a deferred authorization. Therefore the OSAP session's Shared Secret cannot be calculated prior to command execution because the Usage Secret about the data is not known to the authorization engine prior to execution of the command itself. Normally, authorization is done before the command allows execution privilege; deferred authorization is an exception to this rule. Since the TSS that sealed this data in the first place knows this secret, it can authorize the input command that will be authorized using a deferred means within the TPM.

The TPM_UnSeal command checks the type of key, referenced by the Key Handle, to make certain that its type is TCG_KEY_STORAGE. If not, the TPM_UnSeal command generates a command execution error, TCG_INVALID_KEYUSAGE. In addition, the TPM_UnSeal command checks the TCG_KEY_FLAGS to make sure that the key referenced by the Key Handle is not migratable. If so, the command will fail resulting in the error TCG_INVALID_KEYUSAGE. The command also performs integrity checks regarding the encrypted structure of the TCG_STORED_DATA type and returns an error of TCG_NOTSEALED_BLOB if this structure was not constructed correctly. Finally, if the sealed blob is bound to TPM PCR digests, in the form of a composite digest and the host system configuration has been modified relative to sealing, the command errors out with TCG_WRONGPCRVALUE. If any other command errors are returned, check your seal command execution and the input parameters regarding the TPM_UnSeal message block. All of the error conditions described are a direct result of the TPM_Seal command execution, since the only input regarding these checked parameters is in the form of an encrypted TCG_KEY_STORAGE blob.

The output command message block for the TPM_UnSeal is described in Figure 16.12. Simply stated, the output of this command's successful

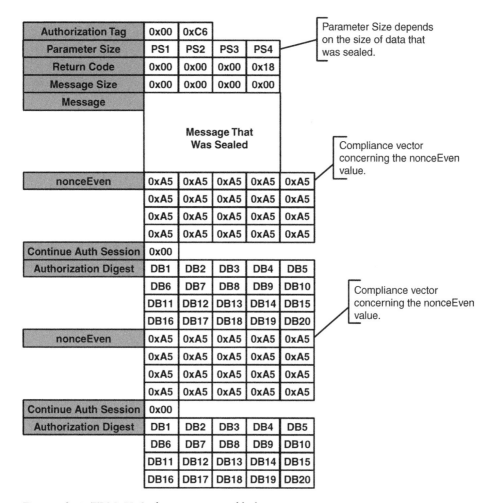

Figure 16.12 TPM_UnSeal output message block

execution returns the secret size along with the secret that was sealed, as well as to the standard passing return code. In addition, this command output is dually authorized and the TSS must validate both the authorization blocks relative to the entities Usage Secret, the encryption key, and the data right-to-know, respectively.

17

The TPM Signature Command

This chapter is concerned with the available commands the Trusted Platform Module (TPM) has to offer concerning RSA signatures. There is only one command available regarding the generic signature of data, outside of specific-purpose signature functionality, defined by Version 1.1b of the TCG Main Specification and it concerns TPM_Sign. Certain TPM vendors provide additional functional support, above and beyond this specified command, but unfortunately, for readers of this book, these commands are proprietary and require signing a Non-Disclosure Agreement (NDA) with the vendor in question. Atmel is no exception.

The rational for this is competitive reasons; if any vendor disclosed its particular specific commands, other vendors would simply add them to their TPM Command Suite. The good news is that Atmel will divulge the nature of each and every vendor-specific command suite, in a form much like the TCG Specifications, with a signed NDA. My recommendation, should you choose to use an Atmel TPM device, is to contact your sales representative and obtain the TPM vendor-specific command documentation.

For embedded design engineers, the Atmel TPM is well-suited, regarding vendor-specific command(s), for support of resource-limited designs and can augment system security design without breaking the bank with regard to system resources. As a matter of fact, the Atmel TPM device will save system resources as opposed to other TPM vendors' devices. While we are on the subject, Atmel is the only TPM device that offers System Management Bus (SMBus) support, so if your system doesn't support the Low Pin Count (LPC) interface, no worries. Not to make this sound like a sales pitch, but the Atmel Corporation is very committed concerning the support of embedded system security design, while the "other guys" are looking for the "big sell".

Atmel designed in a two-wire bus to aid in supporting the TPM within embedded system designs. OK, on to the TPM_Sign command and first things first.

The command input message block concerning the TPM_Sign command is referenced in Figure 17.1. The first three parameters are our three

Authorization Tag	0x00	0xC2			
Parameter Size	PS1	PS2	PS3	PS4	
Ordinal	0x00	0x00	0x00	0x3C	
Key Handle	0x00	0x00	0x00	0x02	
Area to Sign Size	AS1	AS2	AS3	AS4	
Area to Sign	Data				
Authorization Handle	0x00	0x00	0x00	0x00	
nonceOdd	NO1	NO2	NO3	NO4	NO5
	NO6	NO7	NO8	NO9	NO10
	NO11	NO12	NO13	NO14	NO15
	NO16	NO17	NO18	NO19	NO20
Continue Auth Session	0x00				
Authorization Digest	AD1	AD2	AD3	AD4	AD5
	AD6	AD7	AD8	AD9	AD10
	AD11	AD12	AD13	AD14	AD15
	AD16	AD17	AD18	AD19	AD20

Figure 17.1 TPM_Sign input message block

friends: the authorization tag, this is a single authorized command; parameter size; and command ordinal. The reason the command is authorized has to do with the next parameter – the TCG_KEY_HANDLE, a 4-byte value that indexes a TPM internally stored RSA signing key.

Since the RSA key in question, with regard to this example has a usage authorization defined and the AUTH_DATA_USAGE_TYPE is set to the Always condition, the command must be authorized to make use of this key. Previously, we discussed the TPM_LoadKey command, and it is this command that would have been used to load a signing key into the TPM with the previously mentioned attributes. In addition, the RSA key must be defined as either TPM_KEY_SIGN or a TPM_KEY_LEGACY regarding the TCG_KEY_USAGE type. If not, the TPM_Sign command will fail with a TCG_INVALID_KEYUSAGE command return code.

One other characteristic worth mentioning with regard to the signature scheme of the TPM internal key is pointed to by the Key Handle. The three signature schemes are detailed in Chapter 15 about the command TPM_LoadKey. The first choice is really no choice at all, for obvious reasons, and concerns the signature scheme TCG_SS_NONE. The next signature scheme regards the type TCG_SS_RSASSAPKCS1v15_SHA1 and involves the signature of a Secure Hash Algorithm (SHA-1) digest. The final signature scheme is TCG_SS_RSASSAPKCS1v15_DER and this involves the signature of raw data with some data-length limitations. We are going to assume that a signature key was previously loaded into the TPM, pointed to by the Key Handle, and is of the scheme TCG_SS_RSASSAPKCS1v15_SHA1.

The reason for this choice concerns the data size of the message about to be signed, which is unlimited regarding this scheme. Here, the next parameter will always be assigned to a 20-byte signature length because we will be signing a SHA-1 digest. The **areaToSignSize** is therefore assigned the value 0x00000014. The next parameter concerns the message SHA-1 digest itself and will be 20 bytes in length. Hence the reason for the statement that this type of signature scheme can sign a message of any length; the message will always be reduced to 20 bytes by virtue of the SHA-1. For the reasons why the signature scheme TCG_SS_RSASSAPKCS1v15_DER has an **areaToSignSize** limitation, see Chapter 5 about key hierarchy and key management, Section 5.5, Key Cryptographic Algorithm Definition.

The authorization block is attached to the command input message and the authorization digest in this authorization block is related to the TPM internal signing key's Usage Secret. Compile this input message and transmit the command to the TPM. Figure 17.2 describes the resulting command output message block after successful execution.

Authorization Tag	0x00	0xC5			
Parameter Size	PS1	PS2	PS3	PS4	
Return Code	0x00	0x00	0x00	0x00	
Signature Size	AS1	AS2	AS3	AS4	
Signature	Data				
nonceEven	NE1	NE2	NE3	NE4	NE5
	NE6	NE7	NE8	NE9	NE10
	NE11	NE12	NE13	NE14	NE15
	NE16	NE17	NE18	NE19	NE20
Continue Auth Session	0x00				
Authorization Digest	AD1	AD2	AD3	AD4	AD5
	AD6	AD7	AD8	AD9	AD10
	AD11	AD12	AD13	AD14	AD15
	AD16	AD17	AD18	AD19	AD20

Figure 17.2 TPM_Sign output message block

This command output message block contains the expected authorization tag, defined as a single output authorization; parameter size; and the command return code. In addition, there are two output payload parameters, the signature size, and the actual signature blob. Note that the size of the resulting signature is directly related to the size of the keying material used. Therefore 2048-bit, 1024-bit, and 512-bit key sizes result in 256-byte, 128-byte, and 64-byte signature areas, respectively. There is a single authorization block attached to the command output message block, which will be used by the TCG Software Stack (TSS) to validate the output command message relative to the signing keys Usage Secret.

This command can fail in various ways, in addition to the TCG_ INVALID_KEYUSAGE error response defined before. For example, if the **areaToSignSize** is set to zero, the resulting command return code will indicate TCG_BAD_PARAMETER. As always, if this command fails regarding a different error code, check your command input message block for errors.

In summary, the TPM_Sign command must have a key loaded of type TPM_KEY_SIGN or a TPM_KEY_LEGACY regarding the TCG_KEY_USAGE. The TCG_ALG_RSA key algorithm must be defined as either TCG_SS_ RSASSAPKCS1v15_SHA1 or TCG_SS_RSASSAPKCS1v15_DER with regard to the TPM internal key that is about to sign the data. The TCG_SS_ RSASSAPKCS1v15_SHA1 scheme can accommodate any size message, since this scheme signs the digest of the message that is always 20 bytes per SHA-1 algorithm definition. The TCG_SS_RSASSAPKCS1v15_DER scheme can only accommodate a message length for the maximum size of the resulting signature blob minus the distinguished encoding role's (DER) encoding area, which is 11 bytes.

The RNG Command Suite

The Random Number Generation (RNG) Command Suite involves Trusted Platform Module (TPM) commands that support the generation of truly random numbers. I say truly in regard to software-based pseudo-RNG algorithms, which get their seeds from questionable random sources, such as keyboard strokes or some host system metric. The TPM uses hardware-based noise sources that produce random seeds, which are indeterminate regarding the predictability of which seed may be generated next or any cyclic pattern. This hardware-produced seed is then fed into a pseudo-RNG algorithm to output random numbers of varying lengths and convey them to the entity that requested this particular TPM functionality.

The key to any pseudo-RNG regarding randomness is the seed, which is used to start the calculation in motion. Most engineering solutions are based on low-entropy, or stability, characteristics. Introducing greater levels of entropy introduces greater level of system "disorder" or unpredictability. This is precisely the basis of any cryptographic system; introducing levels of high "disorder" or high entropy, which make predictability virtually impossible. The basis of high-entropy systems starts with the random number, and this is why compliance vectors are internal to the TPM that provide predictable results when validating TPM functionality.

Compliance vectors are based on predictable values and when used in the context of an algorithm, such as an RSA encrypt, produce predictable results. The moral of the story is that most engineering solutions are based on low entropy, meaning system behavior is predictable. Cryptographic systems are unlike most engineering realizations – success is based on disorder or high entropy, giving the system a characteristic of unpredictable results, which is what we want.

The good news for embedded system design engineers is that the TPM has the capability to generate quality random numbers and convey these values back to the host system. There is therefore no need for sophisticated random number-generation code, which would take up valuable system resources and may not produce enough entropy to make this solution worth realizing. The host system simply invokes the TPM, via the TPM_ GetRandom command, and the TPM does the rest: supplies the host system with the requested number of random bytes. With this said, let's look at the TPM RNG Command Suite available to the host system.

18.1   The TPM_GetRandom

The TPM_GetRandom command will return the requested number of random data as defined by the input command message block parameter **bytesRequested**. The TPM vendor limits the random value, which is in regard to system resources relative to TPM design specifications. This limit is better conveyed by the TPM vendor and not from the context of this book, but I can safely say that a reasonable number would be in the order of 512 bytes. The TPM_GetRandom command input message block is defined in Figure 18.1.

Authorization Tag	0x00	0xC1		
Parameter Size	0x00	0x00	0x00	0x0C
Ordinal	0x00	0x00	0x00	0x46
Bytes Requested	NB1	NB2	NB3	NB4

Figure 18.1 TPM_GetRandom input message block

This command has a very simple format concerning the usual non-authorized heard information. The only command input payload concerns the number of random bytes to be generated within the given TPM vendor limit. The TPM_GetRandom command output message block is defined in Figure 18.2.

Another simple format concerning this command output message block: the usual output header parameters and two payload parameters, the size of the random data returned, and the random data itself. If the requested number of random bytes is over the limit with regard to the size of the random bytes the TPM can return at once, the maximum limit will be used and the

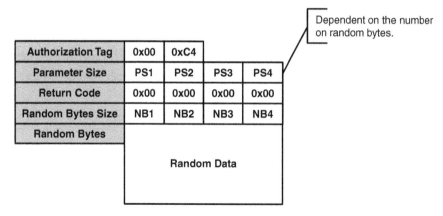

Figure 18.2 TPM_GetRandom output message block

TPM will generate this amount of random data. The next command relative to random number generation is the TPM_StirRandom command.

18.2 The TPM_StirRandom

This command concerns the entropy of the TPM RNG state and a mixing function that adds entropy to the internal random number generator of the TPM. This also applies to the TPM command level outside of the two TPM RNG-specific functions. Remember, the success of any cryptographic system is based on high entropy, and increasing this entropy increases the success of the system as a whole. The input parameters concerning the TPM_StirRandom are defined in Figure 18.3.

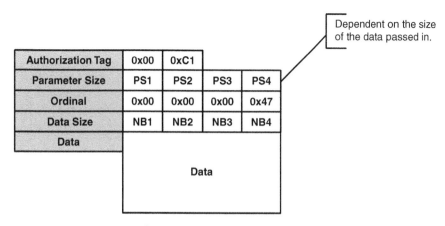

Figure 18.3 TPM_StirRandom input message block

After the header information, the command payload consists of two parameters: the data size and the data, which is used to add the entropy to the RNG. The data size is limited to 255 bytes. The output concerning this command is a simple passed or failed response, as shown in Figure 18.4. The

Authorization Tag	0x00	0xC4		
Parameter Size	0x00	0x00	0x00	0x0A
Return Code	0x00	0x00	0x00	0x00

Figure 18.4 The TPM_StirRandom output message block

command will return a passing result or return the most likely error, **TCG_BAD_DATASIZE**, indicating that the data size contained within the command input message was out of range.

In summary, the RNG-specific TPM commands include TPM_ GetRandom and TPM_StirRandom. The TPM_GetRandom command is used to produce random values with the characteristic of high entropy. The TPM_StirRandom command is used to increase the level of entropy regarding the general TPM RNG, which means not only does the TPM_ GetRandom benefit, but also all cryptographic TPM commands benefit.

19

The PCR Command Suite

This chapter details the Trusted Platform Module (TPM) Command Suite that supports the management of Platform Configuration Registers (PCRs). There are three interesting commands that deal directly with PCRs and support the following functionality: read, modify, and validate or TPM_PcrRead, TPM_Extend, and TPM_Quote. There are other TPM commands that will modify PCR content, but these commands are detailed in Chapter 23, TPM System Deployment Initialization, and involve commands such as TPM_ SHA1CompleteExtend.

The first command investigated here involves the reading of PCR(s) digest(s) in an effort to interrogate specific content; this does not mean that the ability to read translates into the ability to modify. The next command involves the writing or modification of PCR(s) digest(s), more specifically the extension of PCR(s) digest(s), which implies that the resulting digest depends on the previous digest as well as the new digest.

One note: the modification of PCR(s) digest(s) is not very interesting from a security standpoint. The more interesting concept is leveraging the ability of TPM command execution relative to the specific content concerning one or more PCR digests. In addition, sensitive data can also be dependent on the composite digest, which is dependent on selected PCR digest values. The point is that simply modifying PCR digest content does not give the entity performing the modification security privileges regarding TPM command execution or data dependent on previous PCR(s) digest(s) states.

The TPM_Extend command is a support mechanism that allows for the modification of PCR(s) digest(s) in conjunction with valid system configuration updates. Simply modifying the PCR(s) digest(s) in an effort to thwart the host system security only results in limiting TPM functionality, based on keying material and data tied to specific PCR(s) digest(s) that attest to a unique host system configuration. The moral of this story: only use the TPM_Extend and, as we will soon discover, TPM_SHA1CompleteExtend commands if you are completely aware of the host system configuration security architecture. You could limit the ability of the host system, regarding TPM commands that leverage keying material and data bound to previous PCR(s) digest(s), to perform its designed functional requirements.

The final PCR-based command that we will look at involves the validation of one or more PCR(s) digest(s) existing within the TPM at any given point in time. This command pertains to the TPM_Quote; it signs the selected PCR(s) digest(s) in an effort to validate the digest(s) existing within the TPM. The success of this command's execution is dependent on the existence of a valid internal RSA signing key within the TPM at the time of command invocation. This RSA signature key is an integral architectural entity with regard to host system security design that must be a consideration regarding the validation of PCR(s) digest(s) during the lifecycle of the deployed embedded system. For example, any deployed embedded system can be monitored depending on system design considerations, and commanding the TPM_Quote remotely or on-site, and verifying the signature produced by this command can attest to system configuration. If the produced signature does not match the expected value, efforts can commence to correct the situation regarding a potential system modification.

19.1 The TPM_PcrRead

The TPM_PcrRead command will return a 20-byte digest value associated with a PCR index parameter. The following discussion concerning this command execution references compliance vectors, which are associated with predictable PCR digest values. For all the compliance vector PCR digest values, see Chapter 12, Compliance Vectors and Their Purpose.

The TPM_PcrRead command input message block that will return the digest value associated with PCR index 0x00000001 is defined in Figure 19.1.

Authorization Tag	0x00	0xC1		
Parameter Size	0x00	0x00	0x00	0x0E
Ordinal	0x00	0x00	0x00	0x15
PCR Index	0x00	0x00	0x00	0x01

Figure 19.1 TPM_PcrRead (01) input message block

This command input involves the typical input message header and the only command payload parameter concerning the PCR index whose digest is to be returned. The output message block associated with this command execution is defined in Figure 19.2.

Authorization Tag	0x00	0xC1			
Parameter Size	0x00	0x00	0x00	0x1E	
Return Code	0x00	0x00	0x00	0x15	
PCR Digest	0x15	0x8f	0xd1	0x6a	0x35
	0x8f	0x50	0x51	0x2a	0x81
	0x08	0xcf	0xe6	0xec	0xd0
	0xf9	0x07	0xc5	0xc6	0x7c

Figure 19.2 TPM_PcrRead (01) output message block

Again, this is the typical output command header with a single output payload parameter representing the 20-byte PCR digest indexed during command input. If the TPM is in a compliance state, the PCR digest shown in Figure 19.3 will be returned.

PCR 1 Compliance	0x15	0x8f	0xd1	0x6a	0x35
	0x8f	0x50	0x51	0x2a	0x81
	0x08	0xcf	0xe6	0xec	0xd0
	0xf9	0x07	0xc5	0xc6	0x7c

Figure 19.3 Compliance digest representing PCR 1

All PCR digests are assigned predictable values as defined by the Trusted Computing Group (TCG) Compliance Configuration and are labeled PcrA through PcrP – contiguous character assignment – and represent the entire 16 PCR indexing range. This would be a good time to prove to yourself that this command execution results in the same response by sending the preceding defined command to your TPM. One note: this command expects the TPM to be in a post-initialization state and thus expects that the command's TPM_StartUp and TPM_ContinueSelfTest have been successfully executed.

19.2 The TPM_Extend

The TPM_Extend command modifies the 20-byte digest associated with a PCR index. This does not mean that the digest supplied within the input message block of this command will be explicitly written into the shielded memory location pointed to by the PCR index. The Extend function modifies the current PCR digest relative to the input 20-byte digest, as defined in Figure 19.4. Note that this example uses the current compliance vector associated with PCR3 or PcrD, as defined in the TCG Compliance Configuration Specification, and defines the algorithm used to extend this PCR digest.

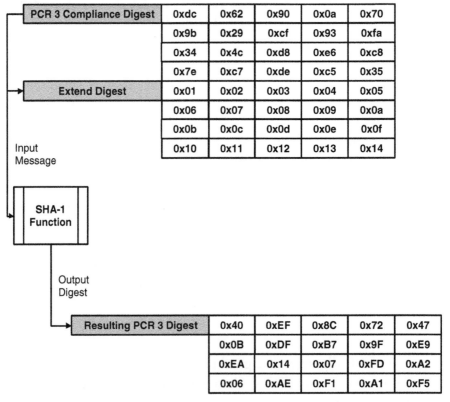

PCR 3 Compliance Digest	0xdc	0x62	0x90	0x0a	0x70
	0x9b	0x29	0xcf	0x93	0xfa
	0x34	0x4c	0xd8	0xe6	0xc8
	0x7e	0xc7	0xde	0xc5	0x35
Extend Digest	0x01	0x02	0x03	0x04	0x05
	0x06	0x07	0x08	0x09	0x0a
	0x0b	0x0c	0x0d	0x0e	0x0f
	0x10	0x11	0x12	0x13	0x14

Input Message

SHA-1 Function

Output Digest

Resulting PCR 3 Digest	0x40	0xEF	0x8C	0x72	0x47
	0x0B	0xDF	0xB7	0x9F	0xE9
	0xEA	0x14	0x07	0xFD	0xA2
	0x06	0xAE	0xF1	0xA1	0xF5

Figure 19.4 Algorithm to extend PCR 3 digest

The input message block for the TPM_Extend command is defined in Figure 19.5 and if you supply the same PCR index and input 20-byte digest, you will obtain the same digest extension described in Figure 19.4.

Authorization Tag	0x00	0xC1			
Parameter Size	0x00	0x00	0x00	0x22	
Ordinal	0x00	0x00	0x00	0x14	
PCR Index	0x00	0x00	0x00	0x03	
Input Digest	0x01	0x02	0x03	0x04	0x05
	0x06	0x07	0x08	0x09	0x0a
	0x0b	0x0c	0x0d	0x0e	0x0f
	0x10	0x11	0x12	0x13	0x14

Figure 19.5 TPM_Extend input message block

To complete the discussion of this command, we must look at the resulting output command message block associated with its invocation as defined in Figure 19.6.

Authorization Tag	0x00	0xC4			
Parameter Size	0x00	0x00	0x00	0x1E	
Return Code	0x00	0x00	0x00	0x00	
Extended Digest	0x40	0xEF	0x8C	0x72	0x47
	0x0B	0xDF	0xB7	0x9F	0xE9
	0xEA	0x14	0x07	0xFD	0xA2
	0x06	0xAE	0xF1	0xA1	0xF5

Figure 19.6 TPM_Extend output message block

If your TPM is in the compliance state and the PCR index along with the input digest are identical to the vectors described in Figure 19.4, you will be able to predict the resulting extended digest. One note: if the TPM is in a disabled state when attempting to execute this command, the PCR digest extension will be facilitated, but the resulting output digest within the output command message block will be all zeros (0x00). The moral of this story: if you want to perform contiguous PCR extensions, for whatever reason, do not attempt to perform this command when the TPM is in a disabled state. The PCR digest will be extended, but you will not get a correct extended digest within the output command message block. If this extension procedure(s) is done correctly, even within a TPM-disabled state, you will be able to predict the digest extension regardless of the digest represented in the command output message. This means, if you enable the TPM and read the newly extended digest indexed by the same PCR reference, you will get back the extended digest value instead of the zero-based default value returned by TPM_Extend while in a disabled state.

19.3 The TPM_Quote

The TPM_Quote command facilitates the validation of specific PCR(s) digest(s) via an RSA signature operation. A prerequisite to the successful execution of this command is the presence – internal to the TPM and indexed by a Key Handle – of an RSA signing key. The command input message block therefore will contain the typical header parameters and three payload parameters: a Key Handle, Anti-replay nonce, and TCG_PCR_SELECTION structure. The Key Handle will index a loaded RSA signing key resident in the TPM, the Anti-replay nonce is a 20-byte random number and the PCR selection structure will identify which PCR(s) digest(s) are to be signed.

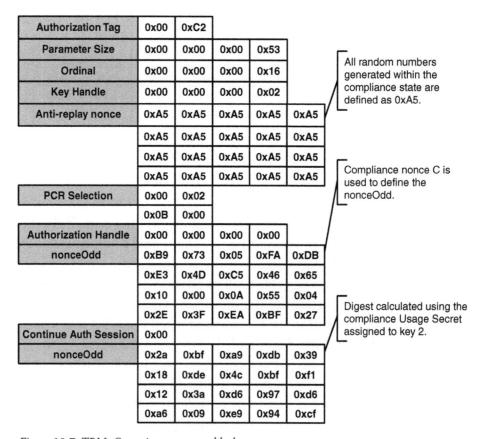

Figure 19.7 TPM_Quote input message block

This input command example follows the compliance vector rules concerning random number generation; that is, all random numbers will contain various length vectors consisting of the value 0xA5. Figure 19.7 describes the TPM_Quote command input message block which will sign PCRs zero, one, and three using the signing key indexed by the Key Handle 0x00000002.

To predict the RSA signature, you must understand how the TPM calculates the area to sign relative to the **TCG_PCR_SELECTION** structure, which is supplied in the command input message block. The quick answer is to simply Secure Hash Algorithm (SHA-1) hash the **TCG_PCR_SELECTION** structure, producing a 20-byte digest and sign this digest. Figure 19.8 defines this procedure relative to the TPM_Quote command input message block in Figure 19.7.

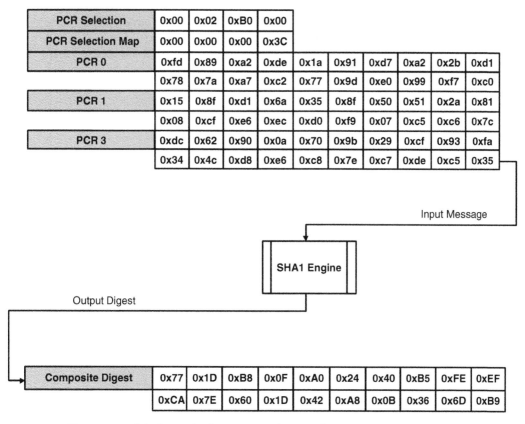

Figure 19.8 Calculating the digest to sign relative to the TCG_PCR_SELECTION structure

Once the TPM receives the TPM_Quote command, the output command message block will contain the input **TCG_PCR_SELECTION** structure, signature size, and signature as payload parameters associated with this

Authorization Tag	0x00	0xC5								
Parameter Size	0x00	0x00	0x01	0x7B						
Return Code	0x00	0x00	0x00	0x00						
PCR Selection	0x00	0x02								
	0x0B	0x00								
PCR Value Field Size	0x00	0x00	0x00	0x3C						
PCR 0 Digest	0xfd	0x89	0xa2	0xde	0x1a	0x91	0xd7	0xa2	0x2b	0xd1
	0x78	0x7a	0xa7	0xc2	0x77	0x9d	0xe0	0x99	0xf7	0xc0
PCR 1 Digest	0x15	0x8f	0xd1	0x6a	0x35	0x8f	0x50	0x51	0x2a	0x81
	0x08	0xcf	0xe6	0xec	0xd0	0xf9	0x07	0xc5	0xc6	0x7c
PCR 3 Digest	0xdc	0x62	0x90	0x0a	0x70	0x9b	0x29	0xcf	0x93	0xfa
	0x34	0x4c	0xd8	0xe6	0xc8	0x7e	0xc7	0xde	0xc5	0x35
Signature Size	0x00	0x00	0x01	0x00						
Signature										

Signature Data

Authorization Handle	0x00	0x00	0x00	0x00	
nonceEven	0xA5	0xA5	0xA5	0xA5	0xA5
	0xA5	0xA5	0xA5	0xA5	0xA5
	0xA5	0xA5	0xA5	0xA5	0xA5
	0xA5	0xA5	0xA5	0xA5	0xA5
Continue Auth Session	0x00				
Authorization Digest	0x2a	0xbf	0xa9	0xdb	0x39
	0x18	0xde	0x4c	0xbf	0xf1
	0x12	0x3a	0xd6	0x97	0xd6
	0xa6	0x09	0xe9	0x94	0xcf

Figure 19.9 TPM_quote output message block

command response. To verify this signature, you can use the corresponding public key associated with the private key used to sign and verify that the **TCG_COMPOSITE_HASH** is correct relative to the PCR selection and digest values contained within the **TCG_PCR_SELECTION** structure. The output command message block associated with the execution of the TPM_Quote is described in Figure 19.9.

One note concerning the PCR selection and the corresponding PCR(s) digest(s): these values must not be defined as zero, meaning that there must be at least one PCR selected and the digest value must not be 20 bytes of zeros. If the PCR selection or any PCR digest is "initialized" to zeros, the TPM_Quote command will fail with a command response of **TCG_INVALID_PCR_INFO**.

There is the concept concerning authorized PCRs and the TCG refers to these types of PCRs as DIRs or Data Integrity Registers. DIRs are similar in use relative to the PCRs with one exception: their use is authorized. Specifically, the writing of digest values to these types of registers is Owner-authorized and the reading of digests contained within the DIRs is non-authorized. For more information regarding DIRs, consult Version 1.1b the TCG Main Specification. In reality, DIRs have not been leveraged to the extent of their corresponding PCR cousins, but this does not mean that you might find a need for these types of registers. Again, consult the TCG Main Specification.

20

The TPM Capability and Self-Test Command Suite

The Trusted Platform Module (TPM) has Command Suites that allow for the interrogation of TPM capabilities and the execution of a full suite of self-tests. *Capability* simply defines the details specific to the TPM this command is being executed on. For example, this command's execution returns to the caller the TPM vendor identification, the algorithms supported by the TPM, the resource's cornering number of keys and authorization sessions supported, vendor-specific commands, and other metrics concerning TPM functionality. In addition to obtaining the TPM capabilities, the caller may also wish that a signing key, internal to the TPM, digitally sign the results. We have already seen the TPM Owner version of this command type and there are two more versions: one for the TPM user to use and one that signs the results in an effort to validate.

Another suite of TPM commands allows for the execution and reporting of the complete TPM self-test, or full suite of TPM self-test. The two commands that facilitate this self-test capability are TPM_SelfTestFull and TPM_GetTestResult. These two explicit self-test invocation commands are defined in addition to the internal self-test protocol that is executed during pre- and post-TPM initialization states.

The TPM_SelfTestFull tests the full capabilities of the TPM and report the failure or success concerning general self-test suite execution. To ascertain which subset of the full self-test suite has failed, in the case of a failed return result from the execution of TPM_SelfTestFull, the command TPM_GetTestResult is invoked. This command is vendor-specific; the output format is defined by the vendor and therefore is talked about in a very general form. Again, the TPM vendor must entertain any vendor-specific discussion and the TPM_GetTestResult is no exception. One of the reasons for this concerns the addition of vendor-specific commands, which must be tested within the context defined by TPM_SelfTestFull and the test results, concerning these commands, must be reported via execution of the TPM_GetTestResult.

The entire point with regard to vendor-specific commands concerns the protection of TPM vendor's IP along with the competitive edge any given vendor-specific command can provide to the individual TPM vendor. For these reasons, a signed Non-Disclosure Agreement (NDA) must protect any TPM vendor-specific command; some TPM vendors may not require this, but Atmel absolutely does. With that said, let's start looking at the suite of TPM capability commands, minus the previously discussed TPM_GetCapabilityOwner.

20.1 The TPM_GetCapability

This command, regarding TPM capability, is much more sophisticated than the owner-based TPM_GetCapabilityOwner. The TPM_GetCapability reports on all the TPM-supported ordinals, including vendor-specific, protocol IDs, algorithms supported, and other metrics that are defined in this chapter. The TPM_GetCapability uses a dual metric to identify the capability to be reported by use of the Capability Area and the Sub-Capability or **CapArea** and **SubCap**. The Capability Area is a fixed 4-byte identifier and the Sub-Capability is a variable identifier dependent on the TPM capability being identified. The command response payload will return the capability response, which is of variable length depending on the reported TPM capability in question. The command input message block is defined in Figure 20.1.

Authorization Tag	0x00	0xC1		
Parameter Size	PS1	PS2	PS3	PS4
Ordinal	0x00	0x00	0x00	0x65
TCG Capability Area	CA1	CA2	CA3	CA4
Sub-Capability Size	SC1	SC2	SC3	SC4
Sub-Capability	Sub-capability information, if sub-capability size is >0			

Figure 20.1 TPM_GetCapability input message block

There are eight types of TPM capabilities that can be queried, defined by Version 1.1b of the Trusted Computing Group (TCG) Main Specification, they

Capability	4-byte Value				Capability Description
TCG_CAP_ORD	0x00	0x00	0x00	0x01	Tests whether a command ordinal is supported.
TCG_CAP_ALG	0x00	0x00	0x00	0x02	Tests whether an algorithm is supported.
TCG_CAP_PID	0x00	0x00	0x00	0x03	Tests whether a protocol is supported.
TCG_CAP_FLAG	0x00	0x00	0x00	0x04	Tests whether a flag is on or off.
TCG_CAP_PROPERTY	0x00	0x00	0x00	0x05	Determines a physical property of the TPM.
TCG_CAP_VERSION	0x00	0x00	0x00	0x06	Gets the current TPM version.
TCG_CAP_KEY_HANDLE	0x00	0x00	0x00	0x07	Gets information regarding Key Handles.
TCG_CAP_CHECK_LOADED	0x00	0x00	0x00	0x08	Gets information about key loading ability.

Figure 20.2 The possible Capability areas regarding TPM_GetCapability

index a specific TPM capability that can be further refined by using the Sub-Capability identifier. Figure 20.2 defines the eight Capability types, which may or may not have Sub-Capabilities.

The first capability identifier defines an ordinal that is to be tested concerning a TPM-supported command. The format is the Capability TCG_CAP_ORD and the Sub-Capability would identify the ordinal to test – for example, TPM_**GetCapability**. Yes we can test the support of this command using this command. Figure 20.3 defines this exact idea by querying the TPM concerning the command support of TPM_GetCapability by executing this command.

TPM_GetCapability Input Message				
Authorization Tag	0x00	0xC1		
Parameter Size	0x00	0x00	0x00	0x16
Ordinal	0x00	0x00	0x00	0x65
TCG Capability Area	0x00	0x00	0x00	0x01
Sub-Capability Size	0x00	0x00	0x00	0x04
Sub-Capability	0x00	0x00	0x00	0x65

TPM_GetCapability Output Message				
Authorization Tag	0x00	0xC4		
Parameter Size	0x00	0x00	0x00	0x0F
Return Code	0x00	0x00	0x00	0x00
Response Size	0x00	0x00	0x00	0x01
Response	0x01			

Figure 20.3 TPM_GetCapability specifically defined to test ordinal support

Notice that the return result is 1 byte, identified by the parameter **respSize** and the actual byte indicating a True or False Boolean regarding ordinal support; in this case, the value is True. Using this variation of the TPM_GetCapability command can test any ordinal; this includes any vendor-specific ordinals.

The next TPM Sub-Capability that can be tested concerning specific algorithm support as defined by the TCG_CAP_ALG capability. The Sub-Capability would simply be the algorithm ID that is to be tested regarding TPM support. Figure 20.4 defines the various algorithms that might be supported by any given TPM, depending on manufacturer.

The input and output command message blocks with regard to this variation of the TPM_GetCapability command is listed in Figure 20.5. Notice that this

Algorithm Identifier	4-byte Value				Algorithm Description
TCG_ALG_RSA	0x00	0x00	0x00	0x01	Supports the RSA algorithm
TCG_ALG_DES	0x00	0x00	0x00	0x02	Supports the DES algorithm
TCG_ALG_3DES	0x00	0x00	0x00	0x03	Supports the triple DES algorithm
TCG_ALG_SHA	0x00	0x00	0x00	0x04	Supports the SHA-1 algorithm
TCG_ALG_HMAC	0x00	0x00	0x00	0x05	Supports the RFC 2104 HMAC algorithm
TCG_ALG_AES	0x00	0x00	0x00	0x06	Supports the AES algorithm

Figure 20.4 TPM_GetCapability and possible algorithm IDs

TPM_GetCapability Input Message				
Authorization Tag	0x00	0xC1		
Parameter Size	0x00	0x00	0x00	0x16
Ordinal	0x00	0x00	0x00	0x65
TCG Capability Area	0x00	0x00	0x00	0x02
Sub-Capability Size	0x00	0x00	0x00	0x04
Sub-Capability	0x00	0x00	0x00	0x01

TPM_GetCapability Output Message				
Authorization Tag	0x00	0xC4		
Parameter Size	0x00	0x00	0x00	0x0F
Return Code	0x00	0x00	0x00	0x00
Response Size	0x00	0x00	0x00	0x001
Response	0x01			

Figure 20.5 TPM_GetCapability specifically defined to test RSA algorithm support

variation is very much like the variation regarding the TPM ordinal support query. The only difference is the Capability, algorithm capability, and the Sub-Capability that defines the algorithm to be tested for support.

Another capability that can be tested concerning TPM support is the support of specific protocols, as defined by TCG_PID. This Capability is identified by TCG_CAP_PID and indicates that the following Sub-Capability will be a refinement concerning the specific protocol ID to be tested with regard to TPM support. Figure 20.6 lists the command input and output messages concerning the testing of the TPM supporting the Object Specific Authorization Protocol (OSAP) ID.

TPM_GetCapability Input Message				
Authorization Tag	0x00	0xC1		
Parameter Size	0x00	0x00	0x00	0x14
Ordinal	0x00	0x00	0x00	0x65
TCG Capability Area	0x00	0x00	0x00	0x03
Sub-Capability Size	0x00	0x00	0x00	0x02
Sub-Capability	0x00	0x02		

TPM_GetCapability Output Message				
Authorization Tag	0x00	0xC4		
Parameter Size	0x00	0x00	0x00	0x0F
Return Code	0x00	0x00	0x00	0x00
Response Size	0x00	0x00	0x00	0x01
Response	0x01			

Figure 20.6 TPM_GetCapability specifically defined to test OSAP support

The TPM manufacturer can be ascertained by use of the TPM_GetCapability command execution with regard to the Capability TCG_CAP_PROPERTY and the Sub-Capability TCG_CAP_PROP_MANUFACTURER. Four TCG_CAP_PROPERTY sub-capabilities are defined in Version 1.1b of the TCG Main Specification; they are listed in Figure 20.7.

Commanding the TPM_GetCapability with the preceding Capability and Sub-Capability parameters can identify the TPM manufacturer. Figure 20.8 lists the input and output messages concerning the identification of the TPM vendor via the TPM_GetCapability command; the vendor is represented by

Sub-Cap for TCG_CAP_PROPERTY	4-byte Value				Sub-Capability Description
TCG_CAP_ORD	0x00	0x00	0x01	0x01	Returns the number of PCRs supported
TCG_CAP_ALG	0x00	0x00	0x01	0x02	Returns the number of DIRs supported
TCG_CAP_PID	0x00	0x00	0x01	0x03	Returns the TPM manufacturer ID
TCG_CAP_FLAG	0x00	0x00	0x01	0x04	Returns the maximum RSA keys that can be loaded

Figure 20.7 TCG_CAP_PROPERTY Sub-Capabilities

TPM_GetCapability Input Message				
Authorization Tag	0x00	0xC1		
Parameter Size	0x00	0x00	0x00	0x16
Ordinal	0x00	0x00	0x00	0x65
TCG Capability Area	0x00	0x00	0x00	0x05
Sub-Capability Size	0x00	0x00	0x00	0x04
Sub-Capability	0x00	0x00	0x01	0x03

TPM_GetCapability Output Message				
Authorization Tag	0x00	0xC4		
Parameter Size	0x00	0x00	0x00	0x12
Return Code	0x00	0x00	0x00	0x00
Response Size	0x00	0x00	0x00	0x04
Response	0x41	0x54	0x4D	0x4C

Figure 20.8 TPM_GetCapability specifically defined to test for TPM manufacturer

four ASCII characters. Any of the four sub-capabilities, regarding the Capability TCG_CAP_PROPERTY, can be tested, as depicted in Figure 20.8, by simply inserting the Sub-Capability of interest.

Another important piece of information concerns the TPM version to an identifier that defines the version of the TCG Specification and the TPM firmware revision. This information is very important when communicating TPM integration issues to the vendor of the device; the TPM vendor may have various versions of firmware deployed and uniquely identifying each of the versions is paramount for aiding customers. The Capability TCG_CAP_VERSION will return the TPM version number and this Capability has no Sub-Capability.

TPM_GetCapability Input Message				
Authorization Tag	0x00	0xC1		
Parameter Size	0x00	0x00	0x00	0x12
Ordinal	0x00	0x00	0x00	0x65
TCG Capability Area	0x00	0x00	0x00	0x06
Sub-Capability Size	0x00	0x00	0x00	0x00

TPM_GetCapability Output Message				
Authorization Tag	0x00	0xC4		
Parameter Size	0x00	0x00	0x00	0x12
Return Code	0x00	0x00	0x00	0x00
Response Size	0x00	0x00	0x00	0x04
Response	0x01	0x01	0x00	0x06

Figure 20.9 TPM_GetCapability specifically defined to test TPM version

Figure 20.9 lists the input and output message blocks concerning this TPM_GetCapability command execution.

Some of the capabilities that can be indexed to help during the execution of applications and the TCG Software Stack (TSS) in regard to dynamic TPM resource allocations. For example, the TSS is about to use a Key Handle, indexing a key that will decrypt a blob it has stored, but the TSS wants to verify that the Key Handle is valid before issuing its use. The TPM_ GetCapability will report the Key Handles that are currently residing within the TPM and the TSS can then look to verify that the Key Handle about to be used is indeed resident in the TPM. Figure 20.10 lists the input and output messages regarding this variation of the TPM_GetCapability command execution.

Finally, what happens if the TSS tries to load a key into the TPM and there are no more resources available to store the keying material? The simple answer is that TPM_LoadKey command execution will fail and the TSS must evict keying material from the TPM before attempting to reload the key in question. Another problem exists that involves key sizes; some keys are bigger than others. What happens if the TSS evicts a smaller key than what it is trying to load into the TPM? Simple, the TPM_LoadKey will fail, once again, with the error concerning TPM resources.

TPM_GetCapability Input Message				
Authorization Tag	0x00	0xC1		
Parameter Size	0x00	0x00	0x00	0x16
Ordinal	0x00	0x00	0x00	0x65
TCG Capability Area	0x00	0x00	0x00	0x07
Sub-Capability Size	0x00	0x00	0x00	0x04
Sub-Capability	0x00	0x00	0x00	0x00

TPM_GetCapability Output Message				
Authorization Tag	0x00	0xC4		
Parameter Size	0x00	0x00	0x00	0x22
Return Code	0x00	0x00	0x00	0x00
Response Size	0x00	0x00	0x00	0x14
Response	0x00	0x00	0x00	0x00
	0x00	0x00	0x00	0x01
	0x00	0x00	0x00	0x02
	0x00	0x00	0x00	0x03
	0x00	0x00	0x00	0x04

Figure 20.10 TPM_GetCapability specifically defined to test Key Handles

There is a simple means to determine whether the key that the TPM wishes to load will be accepted or rejected because of resource issues. The TPM_GetCapability command allows the testing of resources in regard to key loading and available TPM resources based on the algorithm associated with the keying material. Figure 20.11 lists the input and output message blocks used to determine the ability of the TPM to load keying material based on the algorithm and other key descriptors. The return response of the command execution is a simple Yes or No with regard to the ability of the TPM to accommodate the loading of the keying material in question.

In conclusion, the TPM_GetCapability allows for the querying of static and dynamic TPM command and resource data that will define the functional constraints regarding the TPM. In addition, support of vendor-specific commands can be tested for inclusion in the generic TCG Command Suite. When consulting a particular TPM vendor, the TPM_GetCapability command allows the customer to uniquely identify the TPM version for the specific device in question.

TPM_GetCapability Input Message				
Authorization Tag	0x00	0xC1		
Parameter Size	0x00	0x00	0x00	0x2A
Ordinal	0x00	0x00	0x00	0x65
TCG Capability Area	0x00	0x00	0x00	0x08
Sub-Capability Size	0x00	0x00	0x00	0x18
Sub-Capability	0x00	0x00	0x00	0x01
	0x00	0x03		
	0x00	0x01		
	0x00	0x00	0x00	0x0C
	0x00	0x00	0x08	0x00
	0x00	0x00	0x00	0x02
	0x00	0x00	0x00	0x00

TPM_GetCapability Output Message				
Authorization Tag	0x00	0xC4		
Parameter Size	0x00	0x00	0x00	0x0F
Return Code	0x00	0x00	0x00	0x00
Response Size	0x00	0x00	0x00	0x01
Response	0x01			

Figure 20.11 TPM_GetCapability regarding the ability to load a key

20.2 The TPM_GetCapabilitySigned

The TPM Command Suite also includes the ability to sign the capabilities, which are queried via the Capability and Sub-capability indices. The functional aspects of this command as compared to the TPM_GetCapability are identical, with the only difference being the level of certification as provided by the digital signature. There are a few variations in regard to the input message block.

First, the TPM_GetCapabilitySigned, by command declaration, has a Key Handle that points to a signing key and therefore is a single authorized command. In addition, there is an Anti-replay nonce, which is to protect the command input from being used outside of the "place-in-time" when the TPM capabilities where attested. Basically, this command allows for a third party to test the TPM capabilities, attest to the authenticity via an RSA digital signature, and to know that these results are unique or one-to-one in regard to the single instance of command input execution. Figure 20.12 defines the command input message block regarding the TPM_GetCapabilitySigned.

Authorization Tag	0x00	0xC2			
Parameter Size	PS1	PS2	PS3	PS4	
Ordinal	0x00	0x00	0x00	0x64	
Anti-replay nonce	AR1	AR2	AR3	AR4	AR5
	AR6	AR7	AR8	AR9	AR10
	AR11	AR12	AR13	AR14	AR15
	AR16	AR17	AR18	AR19	AR20
TCG Capability Area	CA1	CA2	CA3	CA4	
Sub-Capability Size	SC1	SC2	SC3	SC4	
Sub-Capability	Sub-Capability information, if Sub-Capability size is >0.				
Authorization Handle	AH1	AH2	AH3	AH4	
nonceOdd	NO1	NO2	NO3	NO4	NO5
	NO6	NO7	NO8	NO9	NO10
	NO11	NO12	NO13	NO14	NO15
	NO16	NO17	NO18	NO19	NO20
Continue Auth Session	0x00				
nonceOdd	AD1	AD2	AD3	AD4	AD5
	AD6	AD7	AD8	AD9	AD10
	AD11	AD12	AD13	AD14	AD15
	AD16	AD17	AD18	AD19	AD20

Figure 20.12 TPM_GetCapabilitySigned input message block

This input command leverages the compliance vectors, specifically the signing key indexed by Key Handle 0x00000002. The output message block regarding this commands successful execution is listed in Figure 20.13. The output command message block, in regard to the Capability requested, defines the RSA algorithm as being supported, along with the signature data attesting to the authenticity of this response.

Authorization Tag	0x00	0xC5			
Parameter Size	PS1	PS2	PS3	PS4	
Return Code	0x00	0x00	0x00	0x00	
TCG_Version	TV1	TV2	TV3	TV4	
Response Size	RS1	RS2	RS3	RS4	
Response					
	Response Data				
Signature Size	SS1	SS2	SS3	SS4	
Signature					
	Signature Blob				
nonceEven	NE1	NE2	NE3	NE4	NE5
	NE6	NE7	NE8	NE9	NE10
	NE11	NE12	NE13	NE14	NE15
	NE16	NE17	NE18	NE19	NE20
Continue Auth Session	0x00				
Authorization Digest	AD1	AD2	AD3	AD4	AD5
	AD6	AD7	AD8	AD9	AD10
	AD11	AD12	AD13	AD14	AD15
	AD16	AD17	AD18	AD19	AD20

Figure 20.13 TPM_GetCapabilitySigned output message block

In an embedded system, this command might not be of much use, but the manufacturer could load a signing key into the TPM that could be leveraged to perform this level of certification regarding the TPM capabilities. Normally, a third party is interested in these TPM capabilities and the signing key is actually an Identity Key, which is not covered within the context of this book. Readers who are interested in Identity Key and its use should refer to the TCG Main Specification for further information.

20.3 The TPM_SelfTestFull

The TPM can be commanded to execute a full self-test suite in an effort to validate the integrity of device functionality above and beyond the implicit self-tests performed during the pre- and post-initialization TPM states. This command also is of much more interest to an embedded developer, since the test results can be certified by use of a signing key. Therefore, a deployed system can be commanded to perform a full self-test and the results can be digitally signed and returned to the manufacturer or stored within the system for later retrieval. The results can be simply viewed, in clear without any RSA operation involved, in an effort to test system integrity. If the TPM_SelfTestFull command fails, the TPM will transit to a failure state and maintenance will have to be performed.

First, the TPM must be commanded to perform the full self-test by invoking the input command message block described in Figure 20.14. This figure also defines the output message block associated with a successful command execution.

TPM_SelfTestFull Input Message				
Authorization Tag	0x00	0xC1		
Parameter Size	0x00	0x00	0x00	0x0A
Ordinal	0x00	0x00	0x00	0x50

TPM_SelfTestFull Output Message				
Authorization Tag	0x00	0xC4		
Parameter Size	0x00	0x00	0x00	0x0A
Return Code	0x00	0x00	0x00	0x00

Figure 20.14 TPM_SelfTestFull input and output message blocks

If this command should fail, the return code is anything other than 0x00000000; the TPM will enter the failure state and its functionality will be defined as such. If this is the case, the TPM_GetTestResult may be commanded to ascertain the failure condition existing within the TPM as defined by the TPM_SelfTestFull.

The results defined by the command execution response are TPM vendor-specific; the device manufacturer uses them for diagnostic purposes. The host system developers or reliability engineers simply give the results to the TPM vendor. There is no method for correcting the TPM failure, in most cases, and a new device must be populated within the host system – a good example for migrating keying material.

20.4 The TPM_GetTestResult

This command simply returns the failure code in relation to the TPM in a failure state; the TPM enters a failure state during any self-test failure. This command is very straightforward and since the TPM, in most cases, cannot recover from this state, the execution of this command is for the benefit of the TPM vendor in regard to failure diagnostics. Usually the only means of recovering from this type of TPM failure is to replace the device and send the failed device to the vendor for corrective action. Many times, it benefits the TPM manufacturer and host system designer to understand the causes for individual TPM failures seen in the field – hence the sole purpose of this command. Figure 20.15 defines the input and output command message blocks regarding this command execution.

TPM_SelfTestFull Input Message				
Authorization Tag	0x00	0xC1		
Parameter Size	0x00	0x00	0x00	0x0A
Ordinal	0x00	0x00	0x00	0x54

TPM_SelfTestFull Output Message				
Authorization Tag	0x00	0xC4		
Parameter Size	PS1	PS2	PS3	PS4
Return Code	0x00	0x00	0x00	0x00
Data Size	DS1	DS2	DS3	DS4
Data	Manufacturer-Specific Test Failure Information			

Figure 20.15 TPM_GetTestResult input and output message blocks

20.5 The TPM_CertifySelfTest

This command invokes the TPM full self-test and returns an RSA digital signature on successful self-test completion. If the command fails, the execution of the full self-test fails; the TPM will enter its failure mode state and can be managed by the execution of the TPM_GetTestResult. The purpose of this command is to certify that the TPM is in working order and to not trust an outside entity; however, a digital signature performed by an RSA key, which, in the case of the embedded system, is a key loaded during system deployment.

Again, in regard to PC-based deployments, a third party may be interested in the certification of the results and may use an Identity Key. This command also leverages the use of an Anti-replay nonce, as shown in Figure 20.16.

Authorization Tag	0x00	0xC2			
Parameter Size	PS1	PS2	PS3	PS4	
Ordinal	0x00	0x00	0x00	0x52	
Key Handle	0x00	0x00	0x00	0x02	
Anti-replay nonce	AR1	AR2	AR3	AR4	AR5
	AR6	AR7	AR8	AR9	AR10
	AR11	AR12	AR13	AR14	AR15
	AR16	AR17	AR18	AR19	AR20
Authorization Handle	0x00	0x00	0x00	0x00	
nonceOdd	NO1	NO2	NO3	NO4	NO5
	NO6	NO7	NO8	NO9	NO10
	NO11	NO12	NO13	NO14	NO15
	NO16	NO17	NO18	NO19	NO20
Continue Auth Session	0x00				
Authorization Digest	AD1	AD2	AD3	AD4	AD5
	AD6	AD7	AD8	AD9	AD10
	AD11	AD12	AD13	AD14	AD15
	AD16	AD17	AD18	AD19	AD20

Figure 20.16 TPM_CertifySelfTest input message block

The output of this command, if successful, will be a digital signature attesting to the self-test results as signed by a signature or Identity Key resident in the TPM. Figure 20.17 lists the output command message block in regard to a successful command execution.

Authorization Tag	0x00	0xC5			
Parameter Size	PS1	PS2	PS3	PS4	
Return Code	0x00	0x00	0x00	0x00	
Signature Size	SS1	SS2	SS3	SS4	
Signature					
		Signature Blob			
nonceEven	NE1	NE2	NE3	NE4	NE5
	NE6	NE7	NE8	NE9	NE10
	NE11	NE12	NE13	NE14	NE15
	NE16	NE17	NE18	NE19	NE20
Continue Auth Session	0x00				
Authorization Digest	AD1	AD2	AD3	AD4	AD5
	AD6	AD7	AD8	AD9	AD10
	AD11	AD12	AD13	AD14	AD15
	AD16	AD17	AD18	AD19	AD20

Figure 20.17 TPM_CertifySelfTest output message block

21

The Key Migration and Secret Management Suite

This chapter centers around two Trusted Platform Module (TPM) issues: key migration and user secret management, specifically changing secrets. If you remember, we have seen a variation concerning the modification of established secrets when we discussed the Owner-authorized command, TPM_ChangeAuthOwner. This chapter defines the more generic command regarding entities other than the TPM Owner or Storage Root Key (SRK) that can alter the established secret and takes on the more generic command description of TPM_ChangeAuth. In addition, it looks at the commands that support the migration of keying material internal to the TPM.

These commands help establish a key management functions that is supportive of key backup or key archiving. The migration command execution is built from the Owner-authorized command TPM_AuthorizeMigrationKey as defined in Chapter 14 about Owner-authorized commands. The two migration commands we will look at concern TPM_ConvertMigrationBlob and TPM_CreateMigrationBlob.

21.1   The TPM_CreateMigrationBlob

Migration is normally used to move keys stored in one TPM to another TPM, using a means very similar in nature to executing TPM_CreateWrapKey and TPM_LoadKey. The difference being that the keying material that will be referenced during the TPM_LoadKey command execution exists within another TPM. Therefore, we have two TPMs – TPM_A and TPM_B – and there exists an RSA key pair within TPM_A that we want to load into TPM_B. This concept is defined as *migrating an RSA key pair* from TPM_A to TPM_B and the migration vehicle is an RSA key that already exists in TPM_B.

For example, let's think about how keys are usually loaded into any given TPM – you would leverage the TPM_LoadKey command. What is used to protect the private data associated with the key about to be loaded? A parent key, whose public key provides the encryption mechanism and whose private key is stored in the target TPM. So let's say in TPM_A, the SRK is a parent key to Key_A – the key we wish to migrate to TPM_B. That means that the SRK public key was used to encrypt the private data, concerning Key_A, and the SRK private key was used to decrypt this data and securely load Key_A's private data into TPM_A. Simply put, the SRK public key wrapped Key_A's private data to facilitate the secure loading of this keying material into TPM_A.

Now we want to migrate Key_A from TPM_A to TPM_B and are going to perform a very similar function to the one just described. First, TPM_B has a storage key that we will call Key_B and this key provides a similar protection model, as just described, relative to the SRK and Key_A. So the basic step in migrating Key_A from TPM_A to TPM_B is to "rewrap" Key_A with the public key associated with Key_B stored within TPM_B. The *rewrap* function is performed by first executing the command TPM_AuthorizeMigrationKey, as defined in Chapter 14 about Owner-authorized commands.

Authorizing the migration uses the public key associated with Key_B stored in TPM_B that allows Key_A to be "rewrapped" with Key_B's public key instead of the SRK public key. So the Owner of TPM_A would execute the TPM_AuthorizeMigrationKey using the public key associated with Key_B stored in TPM_B, which facilitates the rewrapping of Key_A stored in TPM_A with Key_B's public key. Once Key_A has been rewrapped, this key can be migrated to TPM_B by executing the command TPM_LoadKey, with the output payload data concerning the execution of TPM_ CreateMigrationBlob within the context of TPM_A. Figure 21.1 defines the concept of rewrapping keys whose parent key was once stored in TPM_A and after rewrapping will be associated with TPM_B.

This rewrapping concept is facilitated by the TPM_CreateMigrationBlob command and is the second command that must be executed during the migration procedure – when migrating a key from one TPM to another. The first step in this procedure concerns the execution of the command TPM_AuthorizeMigrationKey, which must be done in the context of the TPM whose key is to be migrated. The execution of this command creates the **TCG_MIGRATIONKEYAUTH** structure that is an input parameter to TPM_CreateMigrationBlob and is used to authorize the public key that will

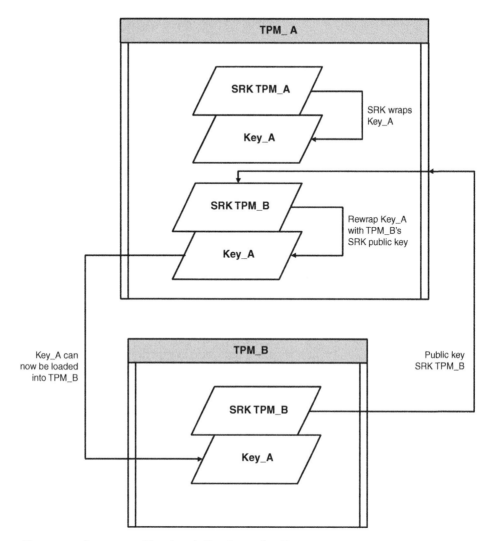

Figure 21.1 Rewrapping Key_A with Key_B stored in TPM_B

perform the reencryption or rewrap. Figure 21.2 shows the command input message block concerning TPM_CreateMigrationBlob.

This command is fairly straightforward when broken down into relative parameters' classifications. First, after the usual header information, we have a Key Handle that indexes the parent key regarding the key we are about to migrate. This key was used to encrypt the private data concerning the key to migrate. The next parameter defines the type of migration we are going to perform; there are two types: a **TCG_MS_MIGRATE** and a **TCG_MS_REWRAP**.

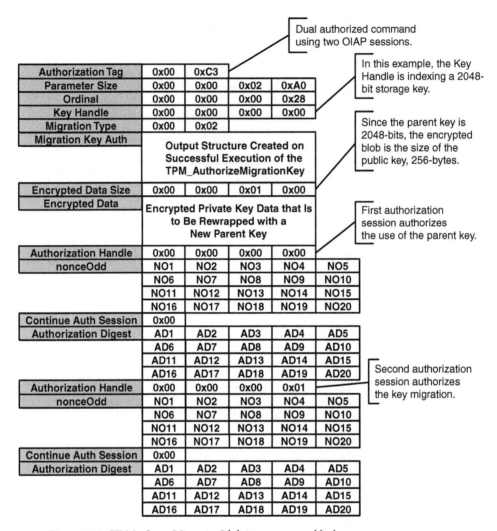

Figure 21.2 TPM_CreateMigrationBlob input message block

The first discussion centers on the **TCG_MS_REWRAP** type of migration. The next parameter involves the RSA public key that we will use to perform the reencryption of the key to migrate. This takes form as the **TCG_MIGRATIONKEYAUTH** structure that is the output payload from a successful TPM_AuthorizeMigrationKey execution. The next two, and last payload parameters, involve the encrypted private key data that is wrapped with the public key indexed by the Key Handle or parent key.

This command is dual authorized and the two entities that need authorization include the parent key (the first authorization session) and the use of the migration public key or migration key authorization (the second authorization session). The second authorization is based on the migration authorization digest associated with the key that is about to be migrated. This is done internally to the TPM by decrypting the encrypted blob that protects the private key data.

One of the protected data parameters is the migration authorization and this is the Authorization Secret concerning the second authorization block. After the command has been authorized, the private data associated with the key to be migrated will be decrypted using the parent's private key, and then encrypted with the public key associated with the migration. When this is done, the output command message block is produced, as defined in Figure 21.3.

Authorization Tag	0x00	0xC6			
Parameter Size	PS1	PS2	PS3	PS4	
Return Code	0x00	0x00	0x00	0x00	
Random Size	0x00	0x00	0x00	0x00	
Encrypted Data Size	0x00	0x00	0x01	0x00	
Encrypted Data	Migrated Key Has Been Rewrapped with the New Parent Public Key and Can Be Used Directly with the Command TPM_LoadKey.				
nonceEven	NE1	NE2	NE3	NE4	NE5
	NE6	NE7	NE8	NE9	NE10
	NE11	NE12	NE13	NE14	NE15
	NE16	NE17	NE18	NE19	NE20
Continue Auth Session	0x00				
Authorization Digest	AD1	AD2	AD3	AD4	AD5
	AD6	AD7	AD8	AD9	AD10
	AD11	AD12	AD13	AD14	AD15
	AD16	AD17	AD18	AD19	AD20
nonceEven	NE1	NE2	NE3	NE4	NE5
	NE6	NE7	NE8	NE9	NE10
	NE11	NE12	NE13	NE14	NE15
	NE16	NE17	NE18	NE19	NE20
Continue Auth Session	0x00				
Authorization Digest	AD1	AD2	AD3	AD4	AD5
	AD6	AD7	AD8	AD9	AD10
	AD11	AD12	AD13	AD14	AD15
	AD16	AD17	AD18	AD19	AD20

Figure 21.3 TPM_CreateMigrationBlob output message block

The typical header parameters are included and the output payload consists of a random size, this value is zero as defined by the **TCG_MS_REWRAP** type, along with the encrypted data size and encrypted data. Note that this encrypted data can be used directly with the input defined by the

TPM_LoadKey command regarding the encrypted private data; you must still populate the remaining command parameters. The TCG Software Stack (TSS) must validate both authorization blocks before using the data provided by this command execution. This is more than likely the most used type of migration; this is especially true when considering the environment defined by most embedded applications. The "backup" mechanism could very well be another TPM-based system used to store deployed keying material that can be recovered or migrated to new TPMs if a deployed TPM fails.

Another type of migration scheme is defined as **TCG_MS_MIGRATE**, which creates a structure that is not directly applicable regarding the TPM_LoadKey command. This mode or type of migration scheme will create a **TCG_MIGRATE_ASMKEY** structure and encrypt this structure with the Owner-authorized migration public key passed during the input command message transfer. The big difference between the **TCG_MS_MIGRATE** and the **TCG_MS_REWRAP** type of migration is that the first type must be "converted" before being loaded into a TPM and the second type can be loaded directly.

The main factor in deciding which type of migration to use depends on what is providing the backup of key archive facilitation. If a third party is providing this service, then the most likely type of migration that will fit this model concerns the **TCG_MS_MIGRATE**. If the backup were being facilitated by a TPM-aware system, then the more efficient method would be the **TCG_MS_REWRAP** type of migration because this type of migration can be used directly by the TPM via the TPM_LoadKey command. If the migration type is defined as **TCG_MS_MIGRATE**, then you must execute the TPM_ConvertMigrationBlob prior to loading the migrated key into another TPM.

21.2   The TPM_ConvertMigrationKey

This command is leveraged if the migration type concerning the execution of TPM_CreateMigrationBlob is **TCG_MS_MIGRATE**. The purpose of this conversion command would be to manipulate the encrypted blob into a form that can be used directly by the TPM_LoadKey command. If the **TCG_MS_MIGRATE** migration type is used during the execution of the TPM_CreateMigrationBlob, the output parameter regarding the random size would not be a zero value and there would be random data included within the output message block for this command. This makes the parameter population very simple, as defined in Figure 21.4.

Authorization Tag	0x00	0xC2			
Parameter Size	PS1	PS2	PS3	PS4	
Ordinal	0x00	0x00	0x00	0x2A	
Key Handle	0x00	0x00	0x00	0x00	
Encrypted Data Size	0x00	0x00	0x01	0x00	
Encrypted Data	XORed and Encrypted Key				
Random Size	RS1	RS2	RS3	RS4	
Encrypted Data	Random Value Used in Data Hiding				
Authorization Handle	0x00	0x00	0x00	0x00	
nonceOdd	NO1	NO2	NO3	NO4	NO5
	NO6	NO7	NO8	NO9	NO10
	NO11	NO12	NO13	NO14	NO15
	NO16	NO17	NO18	NO19	NO20
Continue Auth Session	0x00				
Authorization Digest	AD1	AD2	AD3	AD4	AD5
	AD6	AD7	AD8	AD9	AD10
	AD11	AD12	AD13	AD14	AD15
	AD16	AD17	AD18	AD19	AD20

Figure 21.4 TPM_ConvertMigrationBlob input message block

This input message includes the usual header information and a parent key, whose definition is of the RSA storage key type – a key that can decrypt keying material. This key is protecting the encrypted data and is used to decrypt the blob and has an association concerning the public key used for encryption relative to the TPM_CreateMigrationBlob. In addition, the random number that was generated by the TPM_CreateMigrationBlob command execution is populated into the input parameters defined as **randomSize** and **random**. The command is a single authorized command and is relative to the Usage Secret associated with the Key Handle indexing the keying material inside the TPM.

On successful command execution, the output command message that gets produced is defined in Figure 21.5. The output payload parameter is directly usable by the TPM_LoadKey command in regard to the encrypted portion of the input parameters. The public portion of the TPM_LoadKey input parameters must be populated in addition to this encrypted data blob, which protects the private data concerning the key that is to be loaded.

Authorization Tag	0x00	0xC5			
Parameter Size	PS1	PS2	PS3	PS4	
Return Code	0x00	0x00	0x00	0x00	
Encrypted Data Size	0x00	0x00	0x01	0x00	
Encrypted Data	Encrypted Private Key Data that Can Now Be Used with the TPM_LoadKey Command.				
nonceEven	NE1	NE2	NE3	NE4	NE5
	NE6	NE7	NE8	NE9	NE10
	NE11	NE12	NE13	NE14	NE15
	NE16	NE17	NE18	NE19	NE20
Continue Auth Session	0x00				
Authorization Digest	AD1	AD2	AD3	AD4	AD5
	AD6	AD7	AD8	AD9	AD10
	AD11	AD12	AD13	AD14	AD15
	AD16	AD17	AD18	AD19	AD20

Figure 21.5 TPM_ConvertMigrationBlob output message block

In conclusion, the migration of keying material between two TPMs can be facilitated by the use of the TPM_CreateMigrationBlob. When executing this command, two "modes" of execution can be performed: the TCG_MS_MIGRATE and the TCG_MS_REWRAP. The first mode produces a TCG_MSIGRATE_ASMKEY structure, which must be converted by executing the TPM_ConvertMigrationBlob. The second mode produces an encrypted blob, protecting the private data of the key that is to be migrated; this blob can be used directly by the command TPM_LoadKey.

Determining which type of migration is best for your application depends on the entity that will perform the key backup or archiving function. If this entity has used a TPM-aware system, then it would be best – meaning the least amount of command execution – to use the TCG_MS_REWRAP mode. Otherwise, the TCG_MS_MIGRATE mode can be used, and there will have to be two TPM command executions, the TPM_ConvertMigrationBlob and the TPM_LoadKey, in order to migrate the key into another TPM device.

21.3 The TPM_ChangeAuth

This command is used to change the authorization data that exists within a TCG_STORED_DATA or TCG_KEY structure. These structures are related to the entity types TCG_ET_DATA or TCG_ET_KEY, respectively. This command's discussion focuses on changing the authorization associated with keying material. The same principal is defined for the changing of an authorization associated with data, but since this book deals with the basics, look for this information within Version 1.1b of the TCG Main Specification.

The protocol ID, as defined by the type TCG_PROTOCOL_ID, must be the value defined by TPM_PID_ADCP (see the discussion concerning the command TPM_ChangeAuthOwner). The new authorization – a 20-byte **EncAuth** data blob – is protected by the Shared Secret and algorithm associated with this type of protection. Therefore, the first authorization session associated with the parent handle must be an Object Specific Authorization Protocol (OSAP) session. For discussion regarding the encryption and decryption of the new Authorization Secret, as protected by the **EncAuth**, see Chapter 14 about the command TPM_ChangeAuthOwner. The input command message block associated with this command is defined in Figure 21.6.

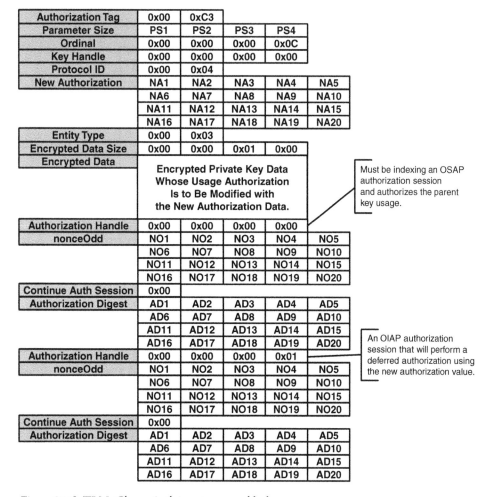

Authorization Tag	0x00	0xC3			
Parameter Size	PS1	PS2	PS3	PS4	
Ordinal	0x00	0x00	0x00	0x0C	
Key Handle	0x00	0x00	0x00	0x00	
Protocol ID	0x00	0x04			
New Authorization	NA1	NA2	NA3	NA4	NA5
	NA6	NA7	NA8	NA9	NA10
	NA11	NA12	NA13	NA14	NA15
	NA16	NA17	NA18	NA19	NA20
Entity Type	0x00	0x03			
Encrypted Data Size	0x00	0x00	0x01	0x00	
Encrypted Data	Encrypted Private Key Data Whose Usage Authorization Is to Be Modified with the New Authorization Data.				
Authorization Handle	0x00	0x00	0x00	0x00	
nonceOdd	NO1	NO2	NO3	NO4	NO5
	NO6	NO7	NO8	NO9	NO10
	NO11	NO12	NO13	NO14	NO15
	NO16	NO17	NO18	NO19	NO20
Continue Auth Session	0x00				
Authorization Digest	AD1	AD2	AD3	AD4	AD5
	AD6	AD7	AD8	AD9	AD10
	AD11	AD12	AD13	AD14	AD15
	AD16	AD17	AD18	AD19	AD20
Authorization Handle	0x00	0x00	0x00	0x01	
nonceOdd	NO1	NO2	NO3	NO4	NO5
	NO6	NO7	NO8	NO9	NO10
	NO11	NO12	NO13	NO14	NO15
	NO16	NO17	NO18	NO19	NO20
Continue Auth Session	0x00				
Authorization Digest	AD1	AD2	AD3	AD4	AD5
	AD6	AD7	AD8	AD9	AD10
	AD11	AD12	AD13	AD14	AD15
	AD16	AD17	AD18	AD19	AD20

Must be indexing an OSAP authorization session and authorizes the parent key usage.

An OIAP authorization session that will perform a deferred authorization using the new authorization value.

Figure 21.6 TPM_ChangeAuth input message block

The **encData** parameters hold the encrypted data defined in either the TCG_STORED_DATA or TCG_KEY structure, respectively, and execution of this command alter the authorization data contained within the encrypted blob. To review the parameters associated with the TCG_KEY structure, see the discussion concerning the TPM_LoadKey command. The TPM, in response to this command invocation decrypts the encrypted blob and the new authorization value and inserts the new authorization value in place of the "old" value in the decrypted structure. When the new authorization value has been replaced, the structure will be encrypted using the keying material indexed by the parent Key Handle. The output payload of this command is simply the encrypted blob that now contains the new authorization data. See Figure 21.7 concerning the output command message block for this command.

Authorization Tag	0x00	0xC6			
Parameter Size	PS1	PS2	PS3	PS4	
Return Code	0x00	0x00	0x00	0x00	
Encrypted Data Size	0x00	0x00	0x01	0x00	
Encrypted Data	Encrypted Private Data with the Authorization Secret Modified per Input Command Parameters.				
nonceEven	NE1	NE2	NE3	NE4	NE5
	NE6	NE7	NE8	NE9	NE10
	NE11	NE12	NE13	NE14	NE15
	NE16	NE17	NE18	NE19	NE20
Continue Auth Session	0x00				
Authorization Digest	AD1	AD2	AD3	AD4	AD5
	AD6	AD7	AD8	AD9	AD10
	AD11	AD12	AD13	AD14	AD15
	AD16	AD17	AD18	AD19	AD20
nonceEven	NE1	NE2	NE3	NE4	NE5
	NE6	NE7	NE8	NE9	NE10
	NE11	NE12	NE13	NE14	NE15
	NE16	NE17	NE18	NE19	NE20
Continue Auth Session	0x00				
Authorization Digest	AD1	AD2	AD3	AD4	AD5
	AD6	AD7	AD8	AD9	AD10
	AD11	AD12	AD13	AD14	AD15
	AD16	AD17	AD18	AD19	AD20

Figure 21.7   TPM_ChangeAuth output message block

The Trusted Device Driver

We have talked about the Command Suite with regard to the Trusted Platform Module (TPM), but this suite's capability is useless without a means to communicate with the TPM, programmatically, with the aid of a well-defined Application Programming Interface (API). Here is where the Trusted Device Driver comes into play; it gives the ability to communicate concerning the TPM physical interface and, logically, through a well-defined driver API. Let's discuss TPM physical communication and the protocol that should be of interest to most embedded system design readers.

At the writing of this book, Atmel produces the only TPM with System Management Bus (SMBus) communication support; all the other flavors of TPM communicate via the Low Pin Count (LPC) interface. The LPC bus is a proprietary protocol that is generally associated with PC motherboard design, hence the popularity of this solution with the "other guys". Atmel, of course, produces an LPC-communication-based TPM, but decided to support the embedded design, with yours truly tasked with designing the logical interface. With that said, I am going to focus on SMBus communication; there are plenty of LPC examples. For one, this book's companion CD includes GPL-licensed Linux driver source code written in C and targeted to the Atmel TPM leveraging the LPC interface. This source will be the "poster-child" for the SMBus-based driver and the only modifications concern replacing the LPC backend with an SMBus backend.

The Trusted Device Driver Library (TDDL) contains the TCG-defined API, which allows any TCG Software Stack (TSS), or "home-brewed", secure stack to gain access to the TPM Command Suite via a well-defined and predictable software interface. Under the hood, the TDDL contains information about how to communicate with specific vendors' TPM and thus be able to abstract this specific knowledge from the calling entity. Later in this chapter, we investigate the API functions and address their usage. The main point to understand is that the embedded system must communicate with the TPM and you have the option to design a tightly coupled realization, with no hope of portability, or to use the TDDL specification to encapsulate the hardware specifics and supply a consistent means to call on TPM commands.

With that said, if you are designing an embedded system that communicates via SMBus, simply remove the "guts" of the supplied TDDL with regard to the LPC bus and replace it with your SMBus interface solution. This means that instead of banging on LPC registers to communicate with the TPM, you will be banging on SMBus registers. The point is that you want to be wiggling the SMBus pins on the TPM, regarding SCL and SDA, and how you do that is not my concern. You could be paying two monkeys, in bananas – one to flip a binary switch for the SCL and likewise for the SDA lines. The real concern regarding this chapter is to understand the definition and use of the API exposed by the TDDL. Now, let's look at each of the API function calls in the services the TDDL supports in regard to TPM command communication.

22.1   The TDDL Interface

Seven different API calls are defined within the TDDL interface, as follows:

1. Tddli_Open
2. Tddli_Close
3. Tddli_Cancel
4. Tddli_GetCapability
5. Tddli_SetCapability
6. Tddli_GetStatus
7. Tddli_TransmitData

Each of the seven API calls supports different information gathering and command transmission in regard to the TPM device the TDDL is communicating with. The TSS software engineer or embedded system design engineer will simply make calls to these defined API functions concerning all communication with the TPM.

In addition, the TDDL can only communicate with a single TPM; the relationship, concerning the backend of the TDDL, is one-to-one. On the other hand, the TSS or "home-brewed" security stack can field multiple applications or embedded "threads" and route any TPM requirements with regard to each process. This is not unlike the idea behind central processing unit (CPU) time-slicing – the TPM services are "shared" between multiple processes running within the embedded system and each process believes that it has the TPM's complete attention. The TSS would be the executive regarding the scheduling of applications requesting TPM services and the TPM itself. With that said, let's look at the seven API calls one can make via the TDDL Interface.

22.2   The Tddli_Open

The API regarding this command call is shown in Figure 22.1. This command API is simple; there are no passed parameters associated with this function. The purpose of this call is to open a communication pipe between the TDDL and the onboard TPM.

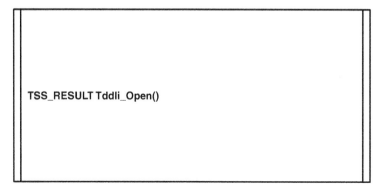

TSS_RESULT Tddli_Open()

Figure 22.1 The Tddli_Open API

As stated, there can only be one association between the TDDL and the TPM, and if the driver has already been opened, the API call will fail with a return response of **TDDL_E_ALREADY_OPENED**. The possible return codes concerning this API call are listed in Figure 22.2.

The command response **TDDL_E_COMPONENT_NOT_FOUND** states that the TPM device was not found; hence the TDDL pipe could not be opened. Good luck if the call returns the very descriptive failure response concerning **TDDL_FAIL**; your only saving grace is that you have control over the driver source!

```
TDDL_SUCCESS
TDDL_E_COMPONENT_NOT_FOUND
TDDL_E_ALREADY_OPENED
TDDL_E_FAIL
```

Figure 22.2 Tddli_Open return responses

```
TSS_RESULT Tddli_Close()
```

Figure 22.3 The Tddli_Close API

22.3   The Tddli_Close

This API call takes no input parameters as well and will attempt to close the communication pipe between the TPM and the TDDL. The API regarding this command is defined in Figure 22.3.

There are three possible response codes that can be supplied after attempted execution concerning the API function; they are defined in Figure 22.4.

Obviously, if you haven't opened the communication pipe you can't close the pipe, hence the return response TDDL_E_ALREADY_CLOSED. Again the descriptive all-informing return response TDDL_FAIL. Once again, see the saving grace.

22.4   The Tddli_Cancel

A useful ability would be to stop the TPM while it's off executing a command, especially if your embedded security stack decides that the command didn't

```
TDDL_SUCCESS
TDDL_E_ALREADY_OPENED
TDDL_E_FAIL
```

Figure 22.4 Tddli_Close return responses

```
TSS_RESULT Tddli_Cancel()
```

Figure 22.5 The Tddli_Cancel API

need to happen or just wants to do it because it can, for whatever reason. The TDDL API function call that will abort the TPM is defined in Figure 22.5.

Man, all of these easy API calls. This API, like the previous two, has no input parameters associated with the function call. Figure 22.6, defines the possible return responses from a call into this API function.

```
TDDL_SUCCESS
TDDL_COMMAND_COMPLETED
TDDL_E_FAIL
```

Figure 22.6 Tddli_Cancel return responses

The return response **TDDL_COMAND_COMPLETED** states that the command being executed by the TPM has completed before the driver issued the abort.

22.5   The **Tddli_GetCapability**

This API is really a wrapper concerning the TPM_GetCapability internal command and will respond with the Capability associated with the **CapArea** and **SubCap** identifiers, which is passed in via the API call. The API defined, for this function, is depicted in Figure 22.7.

The first two parameters define the **CapArea** and **SubCap** identifiers that define what Capability metric will be returned from the TPM. Figure 22.8 defines the possible values these two parameters can be assigned.

```
TSS_RESULT Tddli_GetCapability
(
UINT32    CapArea,
UINT32    SubCap,
BYTE*     pCapBuf,
UINT32*   pCapBufLen
)
```

Figure 22.7 The Tddli_GetCapability API

Cap Area	Sub-Cap Codes	Description
TCG_CAP_VERSION	TSS_CAP_PROP_DRV	Regards the version of the TPM device driver.
TCG_CAP_VERSION	TSS_CAP_PROP_FW	Regards the TPM FW version.
TCG_CAP_VERSION	TSS_CAP_PROP_FW_DATE	Regards the TPM firmware version date.
TCG_CAP_PROPERTY	TSS_CAP_PROP_MANUFACTURER	Regards the name of the TPM vendor.
TCG_CAP_PROPERTY	TSS_CAP_MODULE_TYPE	Regards the vendor-specific designation type of the device.
TCG_CAP_PROPERTY	TSS_CAP_PROP_GLOBAL_STATE	Regards the global state of the module.
TCG_CAP_VENDOR	TSS_CAP_VENDOR_XXX	Regards the vendor-specific capabilities, TPM.

Figure 22.8 The CapArea and SubCap parameter values

If there are any questions regarding the use of these two parameters within this context, please see Chapter 20 about TPM capability.

The next parameter defines a pointer that will refer the data transmitted from the TPM, which is stored within a memory location of the host system. Therefore, before calling this API function, the TSS or embedded security stack would have to allocate some memory block, pointed to by this input parameter. In conjunction with the memory pointer, the last parameter is defined as a read/write buffer length. This means that this pointer will refer a UINT32 defined by the TSS whose value is initialized to the maximum size of the buffer defined by the pointer just described.

This allows the TDDL to determine whether there is enough memory to hold the response within the defined TSS buffer. If not, the TDDL will error the API call. In addition, the TDDL, on successful command completion, will write the response length to the referenced variable defined by the TSS. The possible return responses concerning this API call are defined in Figure 22.9.

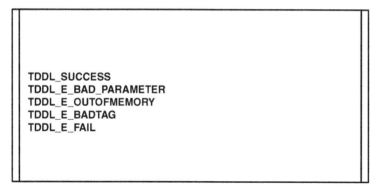

Figure 22.9 Tddli_GetCapability return responses

These responses are specific to the TPM_GetCapability command. The only exception concerns the **TDDL_E_OUTOFMEMORY** return response, which declares that the TSS or embedded security stack – the entity that is presumably making the call to this API – did not allocate enough memory for the response.

22.6 The Tddli_SetCapability

This command is very vendor-specific and no information can be given regarding the **CapArea** and **SubCap** parameters that would be valid within this context. The best source concerning these parameter definitions is the TPM

```
TSS_RESULT Tddli_SetCapability
(
UINT32  CapArea,
UINT32  SubCap,
BYTE*   pSetCapBuf,
UINT32  SetCapBufLen
)
```

Figure 22.10 The Tddli_SetCapability API

```
TSS_RESULT Tddli_GetStatus
(
UINT32  ReqStatusType,
UINT32* pStatus
)
```

Figure 22.11 The Tddli_GetStatus API

vendor and, once obtained, they can be used within the context of this API function call. Figure 22.10 defines the API concerning this function.

22.7 The Tddli_GetStatus

This API function will return the status regarding the TDDL or TPM device, depending on the parameters passed during the call. Figure 22.11 defines this API, which defines two parameters: the **StatusType** and the defined response code.

Figure 22.12 defines the possible status types and the return codes that may be returned as a result. The return results, regarding API functionality, are defined in Figure 22.13.

The error conditions concerning this function call are with regard to the two passed parameters. First, the **StatusType** must be of a type defined in Figure 22.13, and the response code pointer must be defining access to a UINT32.

Status Type	Defined Response Code	Description
TDDL_DRIVER_STATUS	TDDL_DRIVER_OK	TPM driver is operating properly.
TDDL_DRIVER_STATUS	TDDL_DRIVER_FAILED	TPM driver Is not functioning.
TDDL_DRIVER_STATUS	TDDL_DRIVER_NOT_OPENED	The TPM was found, but the driver cannot be opened.
TDDL_DEVICE_STATUS	TDDL_DEVICE_OK	TPM device is operating properly.
TDDL_DEVICE_STATUS	TDDL_DEVICE_UNRECOVERABLE	The TPM error cannot be recovered from.
TDDL_DEVICE_STATUS	TDDL_DEVICE_RECOVERABLE	The TPM error can be recovered from.
TDDL_DEVICE_STATUS	TDDL_DEVICE_NOT_FOUND	TPM device is not present within host system.

Figure 22.12 The status type and defined response code regarding Tddli_GetStatus

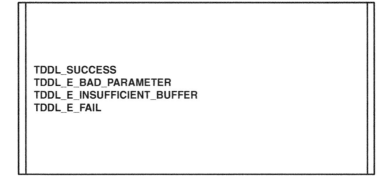

```
TDDL_SUCCESS
TDDL_E_BAD_PARAMETER
TDDL_E_INSUFFICIENT_BUFFER
TDDL_E_FAIL
```

Figure 22.13 Tddli_GetStatus return results

22.8 The Tddli_TransmitData

This is the API function that supports the generic TPM command transmission and receives the TPM response for TSS digestion. The API is straightforward and is defined in Figure 22.14.

This API takes four arguments concerning transmit buffer and buffer size and receive buffer and buffer size. The transmit buffer is fixed by the TSS since the TDDL will not be required to store data in this buffer's boundaries. The receive buffer, on the other hand, must allow the TDDL to declare the size of the TPM response and thus the length parameter associated with this buffer must be passed as a reference to a TSS-defined variable. The return results regarding API functionality are defined in Figure 22.15.

```
TSS_RESULT Tddli_TransmitData
(
BYTE*    pTransmitBuf,
UINT32   TransmitBufLen,
BYTE*    pReceiveBuf,
UINT32*  pReceiveBufLen
)
```

Figure 22.14 The Tddli_TransmitData API

```
TDDL_SUCCESS
TDDL_E_INSUFFICIENT_BUFFER
TDDL_E_IOERROR
TDDL_E_FAIL
```

Figure 22.15 Tddli_TransmitData return response

The TPM command that is to be executed must be populated within the **TransmitBuf** and the command length must be assigned to the **TransmitBufLen** parameter. The receive buffer pointer must be initialized with the maximum size in bytes associated with the receive buffer, as defined by a TSS variable whose reference will be passed.

The Tddli_TransmitData will send the command to the TPM and return the TPM command results via the receive buffer and update the TSS receive buffer length variable with the size of the TPM results length. This command can be used to execute any valid TPM-defined command along with any vendor-specific commands.

TPM System Deployment Initialization

This chapter describes the procedure concerning the preparation of the Trusted Platform Module (TPM) state prior to host system deployment. Some of the considerations involve removing the compliance vectors from the TPM, activating and enabling the TPM, creating a valid Endorsement Key (EK), and configuring the Platform Configuration Register (PCR) digests. This chapter defines the first three issues and any presystem deployment PCR configurations can be addressed relative to the host system and in chapters defining PCR commands.

The reason this book doesn't go into specific PCR configuration detail is because of the variation of possible host system configurations. Readers simply must define the system's metrics that are interesting from a configuration point of view, and exercise the PCR commands that will extend these configuration digests. In addition, the host system bootstrap must be modified to enforce this configuration during system bootup and to protect against configuration alteration during system deployment.

The more pressing matter concerns the state of the TPM, when delivered to the embedded system developer, and the state that the TPM must be in prior to any meaningful functional model. For example, having the compliance EK, as the "root-of-trust" is not a good idea; you must have the vendor of your manufacturing procedure generate a unique EK prior to host system shipment. In addition, developers may not want to design embedded applications that leverage compliance vectors.

One reason for this concerns the management of secrets and keying material; if you have prior knowledge of the data, the management of it can be taken for granted. This simply means to manage data that is known only at time of execution, for example, the TPM_GetPubKey, the application must manage that block of data correctly at that moment in the execution sequence. On the other hand, if prior knowledge of that public key is leveraged, knowingly or unknowingly concerning the application developer, there is a potential data management issue when running in noncompliance mode. Therefore, most of my customers, when I taught TPM system development, chose to develop within the noncompliance state.

The example chapters that follow assume a noncompliance state and detail the procedure regarding data management and the use of a cryptographic support tool. All examples in this chapter and the remaining example-based chapters describe the "C" source, used to obtain the example results, contained on the accompanying CD.

Let's get going with the Manufacturing Initialization Command's sequence. This sequence is the very basic, necessary initialization steps that are generic and can be augmented in regard to specific host system dependencies. When the Manufacturing Initialization has successfully executed, the TPM will be left in a noncompliance state and the only keying material resident in the device will be the EK. The decision of TPM ownership is left to the host system manufacturer; some host systems will have their Owner defined prior to deployment, and others may have the consumer invoke the ownership procedure. The example chapters that follow walk you through TPM ownership command execution in the noncompliance state. The subsection categories here define the execution sequence that will transition the TPM from a compliance state to a normal or deployed state.

Step 1: Execute the TPM_StartUp (Clear) Command

This is your garden-variety TPM start up command that has been discussed throughout this book, but the command message blocks will be inserted here to avoid "page flipping". The TPM_StartUp (clear) input and output message block is defined in Figure 23.1.

Hopefully there are no surprises regarding this command. Remember, the TPM must be "started" prior to any other command execution; if not done, you will always get **TCG_INVALID_POSTINIT** or 0x00000026 error return code. Note, that after executing a successful TPM_StartUp command, if you execute this command again, the same error will be returned; this simply means the command has been previously executed. For those of you scratching your head in response to comparing the TPM_Init and TPM_StartUp

Authorization Tag	0x00	0xC1		
Parameter Size	0x00	0x00	0x00	0x0C
Ordinal	0x00	0x00	0x00	0x99
Startup Type	0x00	0x01		

Authorization Tag	0x00	0xC4		
Parameter Size	0x00	0x00	0x00	0x0A
Return Code	0x00	0x00	0x00	0x00

Figure 23.1 TPM_StartUp (clear) input and output message blocks

commands, the TPM_Init is an alternative or physical method concerning the initialization of the TPM. In reality, most TPM-based system realizations leverage the TPM_StartUp command and your vendor is the best source of information in regard to the TPM_Init command alternative. Once the TPM_StartUp (clear) command has been successfully executed, the next step is to establish physical presence.

Step 2: Execute the TSC_PhysicalPresence Command

TSC_PhysicalPresence command execution establishes physical presence in regard to the TPM device. Now don't get excited, this is a very generic variation of the TPM vendor's specific command execution, and there are much more interesting "modes" concerning this command within a deployed host system realization. To get the full command description and command variations, contact your TPM vendor and get a copy of its TPM Vendor Specific Command Specification. Usually, the vendor specifications follow the Trusted Computing Group (TCG) specifications concerning format and structure definition. TPM physical presence must be established in order to execute the next step in regard to deleting the compliance vectors and state from your TPM; this next step executes the TPM_ForceClear command. But first, we must define the input and output command message blocks concerning the TSC_PhysicalPresence as defined in Figure 23.2.

The execution of this command simply sets the physical presence flag that is internal to the TPM and allows commands that require physical presence to execute – in this case, the TPM_ForceClear command. One comment: the

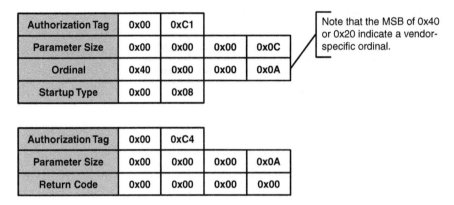

Authorization Tag	0x00	0xC1		
Parameter Size	0x00	0x00	0x00	0x0C
Ordinal	0x40	0x00	0x00	0x0A
Startup Type	0x00	0x08		

Note that the MSB of 0x40 or 0x20 indicate a vendor-specific ordinal.

Authorization Tag	0x00	0xC4		
Parameter Size	0x00	0x00	0x00	0x0A
Return Code	0x00	0x00	0x00	0x00

Figure 23.2 TSC_PhysicalPresence input and output message blocks

TPM_StartUp command execution always sets the physical presence state to False, referencing this generic version of the TSC_PhysicalPresence command. There are variations that will mitigate this condition, but again, discuss this with your TPM vendor.

This sole purpose for the inclusion of the generic physical presence command is to allow readers to follow through on the clearing of compliance vectors and state from their TPM device. Once physical presence has been established, the TPM physical presence state has been set. The next procedural command is to clear the TPM of the compliance vectors and state.

Step 3: Execute the TPM_ForceClear Command

Now we come to the execution sequence that deletes all compliance vectors and state from the TPM; this includes the compliance EK. I mention the EK specifically because there are certain issues that will defined whether the EK is deleted or not. If it is indeed a compliance vector, the EK will be deleted and any attempt to execute the TPM_ReadPubek will fail with a TCG_NO_ENDORSEMENT return code. On the other hand, if the vectors were previously "cleared" and an EK was generated with new keying material loaded into the TPM – "noncompliance" – the TPM_ForceClear execution will delete all keying material with the exception of the EK.

This simply states that the only EK that will be deleted from the TPM concerns the compliance EK. The EK can only be generated once and will never be deleted from the TPM no matter which version of the TPM_ForceClear command is executed. Figure 23.3 details the input and output TPM message block(s) regarding the TPM_ForceClear command execution.

Authorization Tag	0x00	0xC1		
Parameter Size	0x00	0x00	0x00	0x0A
Ordinal	0x40	0x00	0x00	0x5D

Authorization Tag	0x00	0xC4		
Parameter Size	0x00	0x00	0x00	0x0A
Return Code	0x00	0x00	0x00	0x00

Figure 23.3 TPM_ForceClear input and output message blocks

One issue that needs attention concerns the state of the TPM after successful TPM_ForceClear command execution; the TPM is disabled and deactivated. This means that you must enable and activate the TPM before continuing on to the creation of the EK or anything other than the limited command suite(s) defined by these TPM states. Therefore the next two commands to be executed involve the enabling the TPM again and reactivating it.

Step 4: Execute the TPM_PhysicalEnable Command

This command simply enables the TPM and allows command execution outside the suite of commands limited to the TPM-disabled state. As stated previously, the disabled state is a condition that is entered after execution of the TPM_ForceClear command. After executing the TPM_PhysicalEnable command, the internal **TCG_PERSISTANT_FLAGS** – specifically the **disable** flag – will be set to False. Figure 23.4 defines the input and output message blocks associated with this command's execution.

Authorization Tag	0x00	0xC1		
Parameter Size	0x00	0x00	0x00	0x0A
Ordinal	0x40	0x00	0x00	0x6F

Authorization Tag	0x00	0xC4		
Parameter Size	0x00	0x00	0x00	0x0A
Return Code	0x00	0x00	0x00	0x00

Figure 23.4 TPM_PhysicalEnable input and output message blocks

Step 5: Execute the TPM_PhysicalSetDeactivated Command

This command is somewhat dyslexic in nature and we must set the TPM TCG_PERSISTANT_FLAGS – specifically the **deactivated** flag – to False in order to activate the TPM. Therefore the state parameter associated with the command input message block is defined as False and activates the TPM on successful command execution. Figure 23.5 defines the input and output message blocks associated with the successful execution of this command.

Authorization Tag	0x00	0xC1		
Parameter Size	0x00	0x00	0x00	0x0B
Ordinal	0x40	0x00	0x00	0x72
State	0x00			

Authorization Tag	0x00	0xC4		
Parameter Size	0x00	0x00	0x00	0x0A
Return Code	0x00	0x00	0x00	0x00

Figure 23.5 TPM_PhysicalSetDeactivated input and output message blocks

Once the TPM has been activated, the next step is to continue the internal TPM self-test. Note that this command may or may not be necessary, so TPM vendors combine the two separate internal self-test executions in the TPM_StartUp command execution. The execution of the TPM_ContinueSelf Test command does not hinder any TPM state and makes certain that it is in a post-initialization state.

Step 6: Execute the TPM_ContinueSelfTest Command

This command has also been defined previously and is simply instructing the TPM to complete its internal self-test functionality and enter the post-initialization state. Figure 23.6 defines the input and output message blocks for this command.

The final step in the Manufacturing Initialization procedure is to command the TPM to generate the unique and permanent RSA EK pair.

Authorization Tag	0x00	0xC1		
Parameter Size	0x00	0x00	0x00	0x0A
Ordinal	0x40	0x00	0x00	0x53

Authorization Tag	0x00	0xC4		
Parameter Size	0x00	0x00	0x00	0x0A
Return Code	0x00	0x00	0x00	0x00

Figure 23.6 TPM_ContinueSelfTest input and output message blocks

Step 7: Execute the TPM_CreateEndorsementKeyPair Command

This command instructs the TPM to generate a 2048-bit RSA storage key, which is to be identified as the TPM EK. If the EK exists and the TPM is not in compliance state, the EK has already been generated and the TPM will fail the command execution with a TPM_FAIL return code. Remember, the EK can only be generated once – after the TPM compliance state has been cleared – and any attempt to regenerate this key pair will fail. Figure 23.7 defines the input message block concerning this command's execution.

Authorization Tag	0x00	0xC1		
Parameter Size	0x00	0x00	0x00	0x36
Ordinal	0x40	0x00	0x00	0x78
Anti-Replay Nonce	0xEC	0x0C	0xC4	0xD3	0x26
	0xED	0x5F	0xA5	0xF5	0xB1
	0x78	0x1B	0x37	0x5D	0x24
	0xA3	0x36	0x36	0xE5	0x42
TCG_KEY_PARMS	0x00	0x00	0x00	0x01
	0x00	0x03		
	0x00	0x01		
	0x00	0x00	0x00	0x0C
	0x00	0x00	0x08	0x00
	0x00	0x00	0x00	0x02
	0x00	0x00	0x00	0x00

Figure 23.7 TPM_CreateEndorsementKeyPair input message block

Authorization Tag	0x00	0xC4		
Parameter Size	0x00	0x00	0x01	0x3A
Return Code	0x40	0x00	0x00	0x00
TCG_PUBKEY	0x00	0x00	0x00	0x01
	0x00	0x03		
	0x00	0x01		
	0x00	0x00	0x00	0x0C
	0x00	0x00	0x08	0x00
	0x00	0x00	0x00	0x02
	0x00	0x00	0x00	0x00
	0x00	0x00	0x01	0x00
	EK Public Key			
TCG_DIGEST	A 20-byte digest that represents a Checksum of a SHA1 hash whose message is the concatenation of the EK public key and the Anti-replay nonce.			

Figure 23.8 TPM_CreateEndorsementKeyPair output message block

When this command has been successfully executed, the examples defined in later chapters can be addressed using your own TPM device. Figure 23.8 defines the output message block associated with this command's successful execution.

I would suggest that you record the public key information returned on successful command execution; you will need this key to take ownership of the TPM. If you don't record the EK public key at this time, you can always execute the TPM_ReadPubek later. Be aware that there are differences between Versions 1.1 and 1.2 of the TPM regarding the handling of the EK public key – specifically, when or if the TPM_ReadPubek can be executed successfully. More on this difference in the chapter devoted to migrating from the Versions 1.1 to 1.2 of the TPM.

Migrating to Version 1.2 of the TPM

This chapter defines some of the more interesting Trusted Computing Group (TCG) Main Specification modifications in regard to migration from a Trusted Platform Module (TPM) 1.1-based to a 1.2-based host system design. Note that this migration does not isolate any of the Version 1.1-based commands covered in this book but augments and extends its Command Suite. There are major modifications to the Main Specification and an introduction of totally new concepts, in addition to a completely new (Low Pin Count) LPC-based Trusted Interface Specification (TIS). The TIS adds a new concept concerning the physical and logical interface in regard to the LPC bus and introduces the concept of localities or entities that have different priority levels within the TPM device itself. For example, using the TIS interface, the TPM can distinguish between hardware, operating system, and application command requests.

Embedded systems designers will more than likely not be interested in the TIS interface simply because of the extent of system resources that must be designed to support this feature. In personal computer (PC) or Intel x86-based systems, the TIS interface is designed into the chip-set that supports the LPC and other peripheral interface protocols. This makes the task of leveraging the TIS interface much easier when developing stacks, such as the TCG Software Stack (TSS), that are aware of localities. In addition, the most profound difference between the Version 1.1- and 1.2-based TPM realization(s) can be found at the device driver layer and concerns both the physical and the logical aspects based on the TPM LPC interface.

The more interesting TPM device communication interface with regard to embedded designs concerns the two-wire or the System Management Bus

(SMBus) interface; Atmel supports this as well in reference to TPM 1.2-based devices. If you want to know more in regard to the TIS interface and Version 1.2's specific commands, the TCG web site is the best source of information. In addition, your TPM vendor will be able to provide advice about the best design solution given your host system's design requirements.

One difference that I can point out concerning the various options available between the TIS and SMBus is that the SMBus does not support locality distinction. In other words, the SMBus will communicate with the TPM device over a single slave address and thus, because of TPM interface limitations, will not be able to convey which locality is commanding the TPM. This is much like the legacy LPC communication realization (still supported by the 1.2 TPM devices), which communicates with the TPM via a LPC fixed address 0x4E/0x4E and a programmable or filtered LPC address.

This does not mean that the embedded design engineer leveraging SMBus does not have availability to the 1.2 TPM additional Command Suite. This chapter assumes an SMBus TPM interface and defines the Command Suite augmentation along with the TPM state variation in regard to Versions 1.1- to 1.2-based TPM device. Not only does the 1.2 TPM add commands to the available suite, but also some of the entities – for example, the Endorsement Key (EK) – have a different protection profile.

Each of the various aspects that concern the embedded design engineer are addressed in general terms; for command-specific details refer to the specific TCG specification. One reason for this concerns the TCG Main Specification Version 1.2 itself. During this book's writing, the TCG is still making changes to this specification. Again, I strongly recommend going to the TCG web site and downloading the TCG specification(s) associated with Version 1.2 of the TPM. The command structure is identical to that of Version 1.1 of the specification, and the informative comments regarding the TPM and command execution will help you understand the basic TPM protection profile and TPM state before and after commands execute.

24.1 TCG Version 1.2-Based Platform Workgroups

The TPM Working Group

There are several TCG working groups that define different environments in an effort to produce specifications tailored to various levels of security. Each of these security levels, when combined, realizes a trusted platform. The first group that provides the specification that is considered the "root-of-trust" for all of the other working groups concerns the TPM Working Group (TPMWG). This group is responsible for generating the TCG Main Specification, which defines the TPM realization requirements. Most TPM vendors are active within this group in addition to OEMs and central processing unit (CPU) manufacturers. This group works closely with all of the other TCG groups, since the specification generated by this group affects all of the other TCG working groups.

If you are a TCG member and would like to join this group, or any other group for that matter, you simply sign up for membership and join in on the group's weekly call-ins. The main advantage of being a member of the TPMWG concerns the benefit of expressing your views and helping to solve the various issues faced when developing a specification that will be able to provide a guideline for developing a robust TPM. You don't have to be a TPM vendor to make a contribution, but you should have a working knowledge of cryptography at both theoretical and applied levels.

The TSS Working Group

This working group is tasked with the development of a secure stack that will provide the Application Programming Interface (API) used to gain access to the TPM and manage secure information such as keying material and secrets. The secure stack is a very important aspect of the TCG, considering that the group's specification will extend the root-of-trust provided by the TPM to an abstract layer that allows software entities to leverage this trust. In addition, the TSS defines the Trusted Device Driver Layer (TDDL), which describes the well-defined API that all device drivers will leverage to communicate with the TPM physical device. Embedded designs that leverage the SMBus protocol will provide the under layer that supports device communication and gives access to this functionality via the TDDL API. If you are an application Vendor or have a vested interest in defining the API and security issues that are defined by this working group, you might want to consider becoming a member of it. In addition to the TSS specification, this group also defines the header files that define the implementation foundation in regard to this specification.

The Personal Computer Client Working Group

This group is tasked with providing the specification that will define the aspects of integrating the TPM into the PC architecture. This involves PCs' Basic

Input/Output System (BIOS) and boot interface, as well as procedures and other PC-specific aspects that affect security when integrating the TPM into this platform environment. Embedded design engineers can learn a wealth of information by juxtaposing the PC-specific requirements to their own embedded design security issues. Many of the pre-operating system (pre-OS) boot security issues faced within the PC architecture can be migrated to embedded boot-strap requirements, such as Platform Configuration Register (PCR) usage and manipulation.

If you are involved with the PC architecture, especially as BIOS developers and OEMs, and are looking to design in TCG platform security, membership in this group would be a big plus. If you are an embedded design engineer, don't shrug off the group's specification simply because it has PC written all over it, many of the security principles can be directly related to embedded design, especially if you are developing a design for an Intel-based system.

Infrastructure Working Group

This group's main charter is to provide TCG specifications with the ability to transcend into various open platform architectures. There are many enterprise and Internet applications that would like to leverage the TCG platform security models within their specific realizations. This group helps bridge the gap between the TCG specifications and the specific enterprise and Internet infrastructure to allow the TPM root-of-trust to be extended through the various layers of abstraction. If you are involved with Internet or enterprise development solutions, membership in this specific group may be of interest.

The Mobile Working Group

This working group produces the TCG specification that addresses security requirements specific to mobile phones and personal digital assistant (PDA) devices. These devices require specific integration considerations when trying to leverage the TPM, and its protection model within the development environment is defined by this group's target devices.

Other TCG Working Groups

There are other TCG working groups you can join that support various platform and entity realizations that leverage the TPM and software security stacks. These groups are open for membership if you are currently a member of the

TCG. One benefit to group membership is the privilege of participating in the architecture and review of the specifications produced by each group. In addition, you will always have the latest specification(s), defined by that group, well before it is made available to the general public.

24.2   EK Properties

Referring the TCG Main Specification Version 1.1b, the EK pair was created once by successful TPM_CreateEndorsementKeyPair command execution, and this keying material could never be "cleared" or removed from TPM-protected storage. In addition, the EK public key was, by default, made available to all users or TPM entities by successful execution of the TPM_ReadPubek command. The TPM Owner governed the availability of the EK public key, and the owner could protect this information – make the EK public key exclusively available to the TPM Owner – by clearing the persistent flag **readPubek**. This persistent flag is set to the default state of True, meaning that the owner must explicitly clear this flag and limit access to the information exclusively to the TPM Owner entity.

The first deviation of the EK properties just stated concerning migration to a 1.2-based TPM involves access to the EK public key. Referring the 1.1-based TPM, the EK public key is available by default; in contrast, the 1.2-based TPM defaults the TPM persistent flag **readPubek** to a False state. This results in the EK public key being unavailable to any entity other than the TPM Owner after successful TPM_TakeOwnership command execution. The EK public key can be accessed by entities other than the TPM Owner if the owner sets the persistent flag **readPubek** to a True state; however, this is a prerogative of the owner and not a default TPM state. The main concern for this default state change involves the issue of protection of privacy.

Looking at the TPM_ReadPubek command, it is obvious that this command is non-authorized, meaning that any local or remote entity can gain information in regard to the EK public key. If an external entity can gain access to the EK public key, then that external entity may be able to ascertain the identity of the person or entity that this EK public key represents. Remember, the EK is the root-of-trust concerning the TPM and is used to define and attest to valid manufactured TPM devices. Somewhere, there could be a correlation between TPM EK public keys and individuals who bought systems containing the TPM in question. If the EK public key is not protected, by default, an entity could possibly gain knowledge of an individual or entity by correlating this public key information to a list of individuals associated with that cryptographic information.

Note that this scenario is extremely simplified and implies the misuse or mishandling of sensitive data, but why invite opportunity. If the EK public key is made unavailable to external entities, the likelihood of privacy invasion significantly diminishes. Of course, the EK public key must be made available prior to TPM ownership since this data is vital to successful execution of the TPM_TakeOwnership command; the owner and Storage Root Key (SRK) secrets are encrypted via the EK public key.

Another option that is facilitated regarding the Version 1.2-based TPM is the ability to clear or delete the EK from the TPM. This allows the establishment or creation of a new EK RSA key pair, thus allowing the root-of-trust to be based on the new EK. This procedure, outside of simply executing a TPM command, has major security ramifications; all previous cryptographic security has been based on the previous EK. This means that all cryptographic keying material must be associated with the new EK and any certificates – a PC-based issue and maybe for some embedded systems – will have to be "reestablished" so that they will attest to the new EK or root-of-trust. In addition, generating a new EK – a cryptographic vehicle used to attest that it is indeed, for example, an Atmel TPM – might be subject to rigorous physical verification. The bottom line is that "pulling the rug" out from under the TPM's root-of-trust exposes the system to costs and procedures that might not fit the application, which could potentially incur unnecessary costs.

One issue for potential TPM customers – companies that purchase TPMs for their system designs – is the procedure of having the TPM vendor create the EK prior to receiving the devices. This sounds like a great idea, but buyer beware in regard to how the EK data is handled by the TPM vendor. Some vendors use the TPM device itself and command the EK creation, thus protecting the private key associated with this particular keying material. The only key management decision, relative to the TPM vendor, is how to manage the resulting EK public key material. This is the best procedure that TPM customers can hope for and I would certainly recommend this if it were absolutely necessary, relative to the TPM customer, that the EK be generated by the vendor.

The problem with some TPM vendors with regard to EK generation within the production process involves the time factor – how long it takes to get the

TPM through testing profiles. Commanding the TPM to generate its own EK can be a costly process for the TPM vendor, and this opens the door for "alternative" methods to be used to populate the EK within TPM devices. One of these methods is called "squirting" or writing the EK RSA keying material directly into the device instead of waiting for the TPM to create its own RSA key. This saves time since the EK RSA keying material can be generated prior to squirting the data into the TPM, but hence the big problem in regard to the customers for this vendor's TPM. The EK private key is exposed, simply by the nature of this procedure, and the TPM customer must trust the vendor that this keying material has been handled properly.

Considering that the vendor might be processing thousands of devices a month, how can you be sure that there will not be a "slip-up" in the handling of each device's EK private key? The short answer is you can't, and don't ever be "suckered" into thinking that you can. You might pay more to have a "trusted" TPM vendor create the EK within the TPM without ever exposing the EK private key, but the alternative is far too risky. The fact of the matter is that the EK private key is subject to the strength of the vendor's key management system; for those vendors who *squirt* keying material, any security holes could result from the release of the TPM EK RSA key pair to you. Unless you believe that companies don't have disgruntled employees or that every company's internal policies are followed to the letter, especially when production scheduling is concerned, yea right!

The bottom line is control what you can control and don't entrust the root-of-trust to any company that says, "trust me". I would much rather have a TPM vendor add a few cents to cover the additional time to internally create the EK RSA keying material and say "don't trust me". This solution is better for everyone concerned, especially you.

24.3 Extended Context Storage and Restoration

Version 1.2 of the TPM supports a more evolved solution in regard to context storage relative to the Version 1.1 TPM device. For example, if an application is leveraging the TPM and the current state, relative to that application and limited to a state that can be stored, the TPM current state can be saved and restored if the application must give up its TPM thread. The other issue that you must be aware of, especially if you are leveraging the SMBus protocol, is that context *save* and *restore* are sensitive to locality and the SMBus communicates via locality 0 exclusively. This means that the saving and restoration of context information could be diminished when accessing the TPM via the SMBus interface. You should talk with your TPM vendor about limitations concerning the SMBus interface and Version 1.2 of the TPM, specifically the handling of different localities or lack thereof.

24.4 Available Counters

Two types of counter entities have been added to the 1.2-based TPM device relative to the 1.1-based one; these are the Monotonic and Tick Counters. The first, the Monotonic Counter, is supplied to prevent "replay-attacks". It counts indefinitely and allows the TPM, along with the host system, to determine the relative point in time that any TPM command has been executed. Relative time is in regard to the value of the Monotonic Counter on a linear scale and must allow for 7 years at 5-second increments without TPM hardware failure. At a minimum, the TPM will support the establishment of at least four Monotonic Counters and the counter values can be authenticated via an RSA signature. There are five TPM commands associated with the use of Monotonic Counters and the next section describes such commands.

Monotonic Counter

The first TPM command considered here supports the creation of a Monotonic Counter and is named TPM_CreateCounter. The TPM will create the counter, internal to the device, and the "start count" will be assigned by the TPM. This start count is based on an internal monotonic count and the value assigned to this particular counter is the current count plus one. In addition, this command accepts an encrypted authorization that is associated with this counter and its value. The created counter is indexed by a parameter known as the countID and this parameter is used within other Monotonic Counter-specific commands to index a specific counter. This command is Owner authorized and any authorization associated with this particular counter is based on the encrypted authorization secret.

Once a counter has been established, the next logical step concerns the ability to increment the counter indexed by the countID parameter defined by the previous command. Hence the command TPM_Increment Counter, which increments the indexed counter by one. Note that once the counter, indexed by the countID, has been incremented any subsequent counter increments must be associated with this countID, otherwise the TPM will return TPM_BAD_COUNTER. Note that a TPM_StartUp (clear) will void the counter handle; this is equivalent to setting the countID to a NULL value. This command is a single authorized command; the Authorization Secret was defined by the TPM_CreateCounter and established with the encrypted secret.

The TPM_ReadCounter supports the ability to obtain the count in regard to any valid count indexed by the countID. This command is a non-authorized command; knowledge of the count does not provide any security holes, as opposed to changing the value of the counter. In addition, the counter can be released and any attempts to read or increment it – for example, by indexing the associated countID – will fail. The TPM_ReleaseCounter facilitates the "clearing" of the counter identified by the countID. This command is authorized by the secret associated with the particular counter, as defined by the TPM_CreateCounter successful execution. Any sessions, for example, an Object Specific Authorization Protocol (OSAP) authorization, that are tied to this counter will be invalidated due to the fact that the counter will be non-existent. In addition, there is a command that will release a counter, indexed by the countID, that is TPM Owner-authorized.

The Monotonic Counter can be used to prevent replay-attacks associated with the retransmission of TPM commands that have previously been successfully executed. So, for example, if there is a Monotonic Counter value associated with the command being executed and the count is stale – not the count that was expected – command execution will fail. Since the Monotonic Counter increments linearly by one (the setting of it is governed by the TPM and the increment of the counter is authorized), the counter value is protected against tampering and thus can be trusted.

Tick Counter

The Tick Counter is a means of associating a "tick count" with a host system timestamp – the TPM would supply the tick count and the host system would calculate a timestamp. This is no different from getting a timestamp from a PC that makes system time available. The one added advantage that Tick Counters bring to the host system concerns the ability to perform time-stamping within a secure environment. This is facilitated by the TPM supplying the tick count, and the host system manipulating the tick count into a timestamp that is meaningful to this system. Additionally, internal to the TPM, data, in the form of a blob digest, can be "stamped" attesting to the presence of the blob within the TPM at any given moment in time or more specifically a point within a linear count. There are two TPM commands associated with the Tick Counter, as of the current specification reference during this book's penning; they are described next.

The TPM_GetTicks command simply returns the current tick's structure within the return output message block concerning this command. Note that this value is an internal TPM structure representing a linear tick count and the host system would be responsible regarding the correlation between tick count value(s) and absolute or system time. The other Tick Counter-specific command is the TPM_TickStampBlob, which simply signs the Secure Hash Algorithm (SHA-1) digest, representing the blob, with the Tick Counter included. This command simply attests to the time at which any given blob was available or present within the TPM. The command is single authorized and this authorization is based on the entity facilitating the signature operation – in this case, an RSA signing key indexed by the Key Handle in the input message block. Therefore, the Tick Counter simply provides a secure timestamp that can be associated with various blobs (data) and can be used by the host system to correlate this linear count with system time.

24.5   Transport Protection – Encrypted Command(s)

The Transport Command Suite allows for the secure transmission of TPM commands; an entity may not trust the TSS of other middleware between itself and the TPM. Basically, the encryption of the entire command does not produce a meaningful execution flow as such; the data payload associated with the command input and output will be the area bounded for encryption/decryption. There are dual authorizations involved, including the transport command itself, along with the command being transported. All TPMs must support at least one transport session, which are very similar to authorization sessions, making the minimum number of authorization/transport sessions available on any given TPM three. The reason for three is simple, if you want to transport a dual authorized command, the TPM must support three sessions – two for the command authorization and one for the transport session.

Transport sessions have various attributes that can support the modification of the transport execution or the features that are made available by that specific transport session. These attributes, defined by the current TCG specification at the time of this book's writing, are exclusive, logging, and wrapped. The *wrapped* attribute simply states that the transport session will encrypt the command, wrap to payload data, using a symmetric algorithm. The *logging* attribute indicates that transport and command execution will be logged on both input and output command processing, much like the 1.1-based TPM audit functionality. The *exclusive* is a state within the transport command execution that forces the sequence of command execution to be facilitated by that particular transport session. In other words, if a command is executed outside of the protection of the transport session, the session, including any command previously executed and logged within that session, will be null and void. This does not mean that any TPM state, as a result of previous command execution, will be "rolled-back" only that the logging and other secure properties associated with the exclusive transport session will be voided. Again, some of these attribute functional descriptions are subject to change via evolution with regard to the current Version 1.2 of the TCG Main Specification.

There are three TPM commands associated with the establishment, execution, and releasing of transport sessions: TPM_EstablishTransport, TPM_ExecuteTransport, and TPM_ReleaseTransportSigned. The TPM_EstablishTransport command execution is related to the "setting-up" of the transport authorization cache within the TPM, and assigning the session attributes along with the symmetric keying material. There are various states that can be attributed to the established transport session and the wrapping of commands is one option, but it does not have to be enforced. The logging and exclusivity of the established transport session are also optional conditions that are selected during the input message concerning the execution of this command. For example, you could choose to establish a transport session that only performed command logging and didn't wrap any portion of the transported command.

The point is that all or none of the transport attributes can be selected depending on the environment of the host system. Note that setting no transport attribute doesn't make much sense from the TPM perspective; you're simply wrapping the transport-specific command data around another TPM command, but there is nothing stopping you from doing this. The sole purpose of this command is to establish a transport session and give the caller a means of leveraging that session in the form of a transport session handle, no different from an authorization session handle.

Once the transport session has been established, the host system can make use of this session in the means defined by the transport attributes that were defined during the execution of the TPM_EstablishTransport command. If the attribute that defined the session as being wrapped was selected, then the input and output command payload data would be encrypted with the symmetric key. The symmetric key was encrypted and sent to the TPM during the execution associated with transport session establishment. Included within the definition of the TPM_ExecuteTransport command (see the Main Specification Version 1.2) is an algorithm concerning the calculation of Initialization Vector(s) or IV's, which will be used separately during input and output command encryption/decryption.

The transport logging attribute, when set, allows the creation of a digest, which is to be modified during both the transport input and output command message handling, that will allow for an "audit" concerning the sequence of successfully executed transport commands. This logging digest is a SHA-1 hash one based on the population of well-defined structure; it can be reviewed by referring to the TCG Main Specification, Version 1.2. Note that both wrapping and logging transport attributes can be set during the same transport session to allow for the encryption of TPM commands and the logging of command execution.

The final attribute, exclusive, simply defines the TPM state that all following commands executed on that TPM must be associated with that transport session. For example, if you set up a transport session and define the exclusive attribute, all commands must be executed within that transport session. If any command attempts to run within another transport session or on its own, outside of transport protection, the transport session will be nullified. Much more information is available in the TCG Main Specification, and some aspects of this attribute and Transport Command Suite may have changed since the writing of this book. In addition, all three transport attributes can be selected during a single transport session establishment.

The main goal of the transport session is to establish a secure portal concerning the execution of TPM commands that might be executing via unsecured communication signals. In addition, the application may not trust the TSS providing access to the TPM, and may want to hide command execution details for this entity. Also, a secure application or OS may want to establish a secure thread in which commands can only be handled within that thread and any attempts to circumvent this security feature will result in the closing and detection of the secure pipe to the TPM device.

24.6 Summary

The following are just some of the highlights concerning migration from a Version 1.1- to a 1.2-based TPM. One of the biggest benefits when migrating to the 1.2 TPM device, as far as embedded systems are concerned, is the protection facilitated by the transport-based commands. The commands can be encrypted and provide another layer of protection against the "sniffing" of the communication bus or wires during deployed system execution. This cryptographic protection makes the task of reverse engineering much more difficult. Contact your TPM vendor for more information concerning Version 1.2's TPM functional details, especially if you are leveraging the two-wire interface as opposed to the standard LPC interface.

There are many more details not discussed within the confines of this book regarding specific 1.2 TPM commands, but a thorough understanding of Version 1.1 of the TPM will provide a solid foundation concerning the understanding of 1.2 TPM-specific commands. All of the commands discussed in this book are applicable to the Version 1.2 of the TPM; arguably they may be deprecated, but functional. The important issues concerning command compilation and authorization are still applicable and outline some of the more difficult TPM concepts. The 1.2 Main Specifications are different from Version 1.1 in the fact that the document has been separated into three specific categories: command, structure, and information discussion documents. This allows for a more detailed and informative discussion in regard to reading this specification and getting conceptual information regarding the TPM and its command support. The command format remains the same and all of the knowledge gained from study and development relative to the Version 1.1 TPM can be applied, making the migration to the 1.2 TPM cumulative. In fact, most TPM vendors (I know of a few today) will only make available the 1.2 TPM, since all 1.1-based commands can be leveraged with this device without ever "turning on" specific 1.2 TPM command-level support.

One other feature that the 1.2 TPM offers is a more generic and robust Nevada (NV) Store functionality. This is different from Version 1.1 with regard to the strictness of the type, or structure of, data that can be stored within the TPM's physical boundary. For example, some users of the TPM want certificate-related data stored within the TPM so that when they want to verify an X509 Certificate bound to the TPM, they can use this stored data to facilitate that goal. Embedded system designers can use this storage space to facilitate the secure storage of data relative to their particular target solution.

The bottom line is that the Version 1.2 TPM offers a much improved NV Storage capability that is pliable relative to the host system. Your TPM vendor will provide you with a wealth of knowledge regarding this very helpful aspect for the integration of the TPM within your embedded system.

Example One: TPM Ownership

This chapter deals with the command specifics regarding the steps defined when establishing Trusted Platform Module (TPM) ownership. This example will be facilitated by the compliance vectors stored in the TPM at the time of device delivery from the distributor or vendor and existing in the TPM until a TPM_ForceClear command is issued. The reason behind this logic is to allow predictable results from the example's command execution and thus instill some level of confidence during the execution of them.

The procedure for taking ownership and any other compliance-based command is the same regardless of TPM state concerning compliance or "normal" state; the basic difference concerns predictable results. One concern that readers should be aware of entails the Optimal Asymmetric Encryption Padding (OAEP) encoding and the use of nonrandom padding so as to allow predictable encoded data prior to RSA encryption. The best discussion concerning this particular issue is in the Trusted Computing Group (TCG) Compliance Specification; another resource is a TPM vendor.

All of the following examples will leverage the predictable data values resulting from fixed Object Specific Authorization Protocol (OSAP) encoding and RSA keys. If you perform a TPM_ForceClear, the procedure with regard to the TPM command execution will be the same; only the randomness, or entropy, of the data will be of concern.

Step 1: Generation and Encrypting the Secrets

Before compiling the TPM_TakeOwnership command input message block, there are a few steps that should be performed that involve the Endorsement Key (EK) public key and the encryption of entity secrets. These secrets are specifically the Storage Root Key (SRK) and Owner-authorization secrets that will be leveraged during command authorization specifically referencing these entities. With that said, we must decide on the authorization secrets that we want associated with the SRK and Owner entities.

In this example, I chose some very well-known or predictable values, for example purposes only; you should choose randomly generated values when system deployment is addressed. The 20-byte Authorization Secrets that I will use as examples are shown in Figure 25.1. The values that I have chosen

Owner Secret									
A2	18	08	B2	18	38	A2	06	50	16
E6	E2	00	5C	CD	5C	80	3F	25	B9

SRK Secret									
36	F7	56	60	E9	73	F8	8A	C0	80
CD	95	65	C0	3D	5D	8F	AA	CE	94

Figure 25.1 Example SRK and Owner Authorization Secrets

are defined in the TCG Compliance Configuration Specification under the heading "Nonces, secrets".

Once we have decided on the SRK and Owner Secrets, the next step is to encrypt these values with the EK public key in anticipation of building a complete TPM_TakeOwnership command input message block. One thing that I want to mention concerns the tools available concerning public key cryptography or, more specifically, the cryptographic support in regard to encrypting and verifying blobs using RSA public keying material. Remember that the TPM deals with RSA private keys, specifically; it protects the keys and uses private keys to perform private RSA key cryptographic operations. Therefore, you must support public key cryptographic algorithms outside of the TPM – for example, RSA encryption and signature verification.

There are open-source tools available that you can leverage, but most tools support much more than an RSA cryptographic engine and can have a very large memory footprint if used in their entirety. OpenSSL and cryptLib are two examples of open-source cryptographic tools that you can leverage to support the public key cryptographic functions needed, but your vendor is more than likely the best source for information regarding this concern. During the example development and integration, I am using the Atmel vendor-specific support tools with regard to RSA public key cryptographic algorithms; if you're targeting the Atmel TPM, give customer support a call. With that said, let's encrypt the SRK and Owner Secrets that will be included in the input message block of the TPM_TakeOwnership command.

First, the EK is a 2048-bit key and the resulting encrypted blob will be 256 bytes with regard to the two secrets that are to be encrypted. Next, the encoding scheme to be, leveraged during encoding of the secret(s) are defined as **TCG_ES_RSAESOAEP_SHA1_MGF** and your public key encryption engine must support this method of encoding. Next, we must get the EK public key data; this can be accomplished by successful execution of the TPM_ReadPubek command (see Chapter 13 in regard to the input command message). Once you have the EK public key it will be set as the encryption key for the remainder of this example.

Figure 25.2 redefines the EK public compliance key that will be returned if your TPM is within the TCG compliance state. If your public key is not identical to this RSA public key, your TPM is not in compliance state, but you can follow this example. The only difference between the compliance

0xab	0x56	0x7c	0x0e	0x60	0x8c	0x5c	0x18	0x9e	0x90	0x2c	0x37	0x32	0xcf	0xe3	0xfe
0x4f	0xa7	0xb5	0x0c	0x78	0xa1	0x5d	0xa7	0x39	0xeb	0xc0	0x06	0x87	0x05	0xdb	0x1f
0xe4	0xab	0x2a	0x9a	0x68	0xe3	0x5b	0xb6	0xfb	0x27	0x69	0x5a	0x4b	0xe2	0x90	0x65
0x04	0xb2	0x78	0xcf	0x44	0x02	0x7c	0x16	0x4c	0xfb	0xf5	0xf0	0xf6	0x25	0x7d	0x31
0xf1	0x2e	0xd8	0x67	0x93	0x5a	0x48	0xb2	0xc1	0x4c	0x16	0xfd	0x97	0xe5	0x86	0x65
0x4a	0x2e	0x07	0x4b	0x14	0x78	0xf7	0x66	0x83	0x66	0x05	0xb0	0xea	0xec	0x1e	0x16
0xcf	0xf9	0xf9	0xc5	0x5c	0xbc	0x7b	0x42	0x24	0xa1	0xa7	0x1b	0x55	0xd7	0x4b	0xb1
0x62	0x7f	0x90	0x88	0xee	0xfb	0xfb	0x26	0xb1	0x4f	0x56	0x97	0x8c	0xd0	0x12	0x05
0xa6	0xef	0x09	0xc9	0x08	0x10	0xf2	0x1b	0x65	0x9c	0xf2	0x05	0x7b	0xcc	0x4e	0x6a
0x65	0x0c	0x1c	0xe1	0xb5	0x3e	0x86	0x7d	0xf8	0x0b	0x8b	0x6f	0xe3	0x72	0x2b	0xcb
0xc9	0x3d	0xf8	0x61	0xf4	0x83	0x74	0xb1	0x38	0xa6	0xce	0xde	0x18	0x7f	0x8d	0xc4
0x8f	0xa1	0x8e	0xa6	0xac	0x71	0xa4	0x89	0x60	0xd3	0x3e	0x5f	0x3d	0x18	0x5c	0x32
0x6c	0x96	0x1d	0x84	0x8b	0x50	0xc3	0x5b	0x68	0x5c	0x16	0x2d	0x9c	0xbb	0xf1	0x79
0x60	0x6e	0xc9	0x25	0xaa	0xec	0x26	0x9e	0x9e	0xd4	0xd6	0x89	0xf3	0xff	0x23	0xaa
0x75	0x46	0x3b	0x4a	0xea	0x1d	0xe5	0x03	0xb9	0xac	0x6d	0xf8	0x2d	0x88	0xff	0x84
0x12	0xb8	0x47	0xcf	0x3a	0x32	0xc9	0x66	0xc6	0xe3	0x2c	0x1f	0x7d	0x30	0xd8	0x99

Figure 25.2 Compliance EK public key data

state and "normal" state is in regard to the predictability of the resulting encrypted blob; the procedure is the same no matter which state your TPM is in.

Now we supply the public key and the data to be encrypted; in this case, it will be the first secret or the Owner authorization 20-byte secret defined in Figure 25.1. The output of the encryption operation will be a 256-byte blob as shown in Figure 25.3.

Next, we perform the same encryption operation; only this time, we encrypt the SRK Authorization Secret using the same EK public key. The resulting output of the encryption operation will, once again, be a 256-byte blob as shown in Figure 25.4.

```
59 4e c5 e2 6b a0 55 79 de 7c 8e ac 9d 1b 7f 59
4f f5 74 b9 5a d1 f9 f7 0d 3d 91 7c eb 72 e0 c3
c9 ce 13 d9 ad 8b 10 d9 b1 61 2f 60 63 07 cd 60
2d d2 86 2e fc 08 24 4c 9c f1 17 50 0c db 90 3c
de db 4b 53 83 c0 ca 59 a8 19 3d 45 81 52 a7 eb
32 23 c1 61 5e b3 9e 4b b8 17 4f 66 9a 2b 0b f1
8b e4 02 31 2e a6 51 3f 99 73 8d 16 59 08 45 51
e7 24 63 b3 57 bc 79 56 38 ae 21 21 e9 d9 13 84
18 fe 43 09 1f 77 be af 92 d4 f4 c2 3b d8 cf 0e
0c 96 37 3f 95 2b 00 18 1c 18 a7 a0 05 53 f3 78
b1 10 86 ea 32 33 5f a3 a5 d0 0e 55 f3 2d fb 37
c5 34 55 8d 55 d4 eb a7 91 37 44 37 5b 16 a1 fd
3b aa 5b 02 cd d9 9d 85 2b 39 96 c5 9c 35 f8 2c
65 7d d0 c7 1a c7 94 d7 35 74 33 9e 77 c6 a4 d2
e2 72 c8 99 6d 71 b6 fb f7 2d 7c 27 ef 92 e4 35
66 82 d9 20 62 2e c7 c1 1e ae fa c2 b8 ba 8f 35
```

Figure 25.3 The resulting Owner Authorization Secret, EK encrypted

```
00 22 7d 62 5c 3f d5 eb 13 ed 66 b9 66 22 c5 ea
e2 f0 07 45 a7 27 2a cc a4 85 2f e3 41 e7 b4 fd
04 87 83 f1 b8 2a a5 1a 0c 20 97 19 2a 58 c5 59
ed dd 98 cc fe c8 40 46 ff a2 03 0f ac b8 83 a0
91 27 8c 48 0f 74 95 16 96 9b 3e 02 1e b3 7e 65
c7 fd bc fc 71 b7 39 d9 b0 b3 2e ce 1d c6 ab cd
9a 08 48 d3 cf 0d cf 80 37 ec 3d 18 ef 4f 95 14
b0 1d ea 01 92 fa b8 25 c4 c0 d7 5d 6f e6 a2 87
c9 05 bc 5d 75 42 9d c5 33 b5 9c cb fb ed 1f 65
cb 47 05 08 c9 06 d3 16 c5 da 8c d4 b3 4e 0f 92
3b 24 a3 dd 6c 65 e0 c4 e5 d7 db 28 7a df d0 96
21 45 af 23 86 4a 0b 4f 70 59 75 d7 fc dd d0 bf
02 7f 92 5e e9 27 1b 6e 08 64 18 ef 6d 19 94 d5
81 c2 80 58 53 72 05 f1 cc f0 cd 75 e7 51 13 1c
ee d5 d9 87 99 21 86 5a 82 b6 65 28 71 85 e6 83
6a c3 08 41 b3 17 c7 0e 35 86 dc d1 b5 68 46 c3
```

Figure 25.4 The resulting SRK Authorization Secret, EK encrypted

We will save these two 256-byte cryptographic blobs to use later when we compile the TPM_TakeOwnership command input message block. In addition, remember your Owner authorization 20-byte secret, the one we just encrypted with the EK public key; you will need this to authorize the TPM_TakeOwnership command.

Step 2: Populating the TCG_KEY Structure

The other big command parameter that must be dealt with concerns the information that will be supplied in regard to the type of key that will be generated representing the SRK RSA key pair. This is important because the SRK must be a specific type of TCG key, along with being of the RSA key type; any errors in describing this key will result in the failure of the TPM_TakeOwnership command. For more information about the exact parameter population in regard to this command, see Chapter 13. Figure 25.5 depicts the populated TCG key structure that will be included within the command input message block in regard to TPM_TakeOwnership command execution.

The TPM version is variable depending on your device and when you received it. Make sure you execute a TPM_GetCapability command to ascertain your version identifier.

TCG_KEY	0x01	0x01	0x00	0x01
	0x00	0x11		
	0x00	0x00	0x00	0x00
	0x01			
	0x00	0x00	0x00	0x00
	0x00	0x03		
	0x00	0x01		
	0x00	0x00	0x00	0x0C
	0x00	0x00	0x08	0x00
	0x00	0x00	0x00	0x02
	0x00	0x00	0x00	0x00
	0x00	0x00	0x00	0x00
	0x00	0x00	0x00	0x00
	0x00	0x00	0x00	0x00

Figure 25.5 The TCG key structure relating to TPM_TakeOwnership

The only remaining payload parameter that has not been discussed in this example chapter is the protocol ID. This parameter must be set to the TCG value TCG_PID_OWNER, as described in Chapter 13. When all of the previously mentioned parameters are defined, the next step is to compile each parameter into a complete TPM_TakeOwnership command input message. This includes the typical header information: authorization tag, parameter size, and command ordinal. Figure 25.6 depicts this parameter ordering.

Authorization Tag	0x00	0xC2		
Parameter Size	0x00	0x00	0x02	0x70
Ordinal	0x00	0x00	0x00	0x0D
TCG_PROTOCOL ID	0x00	0x05		
Size of Enc Owner Secret	0x00	0x00	0x01	0x00
Encrypted Owner Secret	Insert Data From Figure 25.3 Representing the Owner-Encrypted Secret.			
Size of Enc SRK Secret	0x00	0x00	0x01	0x00
Encrypted SRK Secret	Insert Data From Figure 25.4 Representing the SRK-Encrypted Data.			
TCG_KEY	0x01	0x01	0x00	0x01
	0x00	0x11		
	0x00	0x00	0x00	0x00
	0x01			
	0x00	0x00	0x00	0x01
	0x00	0x03		
	0x00	0x01		
	0x00	0x00	0x00	0x0C
	0x00	0x00	0x08	0x00
	0x00	0x00	0x00	0x02
	0x00	0x00	0x00	0x00
	0x00	0x00	0x00	0x00
	0x00	0x00	0x00	0x00
	0x00	0x00	0x00	0x00

Figure 25.6 TPM_TakeOwnership input message block

There should be no surprises here; notice that I left out the information regarding the command parameter size. We have one more step to perform: compile the authorization block for this command. Once the authorization block has been established, the command length will be the size of the input message plus the authorization block size. If you would rather calculate the parameter size prior to compiling the authorization block, simply add 45 bytes to the input command message length. The TPM_TakeOwnership is a single authorized command and thus has only one input authorization block.

Step 3: Calculating the Input Authorization Digest

The authorization digest is one of the most critical steps concerning TPM command compilation. Be careful to follow the specific rules defined for hashing input/output command payload data and final command authorization digest(s). I would suggest that you develop a generic authorization engine prior to integrating multiple TPM command support. You might want to start with the "typical" or straightforward command authorization calculations and then modify the authorization engine with regard to exceptions to the rules. During the "early days", I was given the task of developing the authorization engine for the TPM firmware; I can tell you from experience that you must get this functional block accurate or suffer the consequences.

What I mean by this is that all commands go through the authorization engine prior to specific command functionality, and if your authorization engine is not performing as specified, this will put you in the critical path concerning every command development. This is a place that you do not want to be and fully testing each permutation, with regard to authorization typical and atypical calculations, will keep you out of that critical path.

Figure 25.7 defines the complete command input message block concerning the TPM_TakeOwnership command. The payload data that will be Secure Hash Algorithm (SHA-1) hashed is shaded and is to be the input message to the hash algorithm.

If your TPM is in the compliance state and your encrypted blobs are identical to the example data, which they should be given the predictable OAEP encoding, RSA public key, and secret information, your payload digest should match the examples. Notice that this command does not include any Key Handle information and as such allows for a contiguous input message to the SHA-1 engine. If the command you are authorizing does have Key Handles in the input message block, you must arrange the payload data – make this data contiguous – by removing the Key Handle data and adjusting the payload information. Figure 25.7, in addition to defining the SHA-1 message, defines the 20-byte payload digest that is to be used as the first Hash-based Message Authentication Code (HMAC) parameter when calculating the authorization digest.

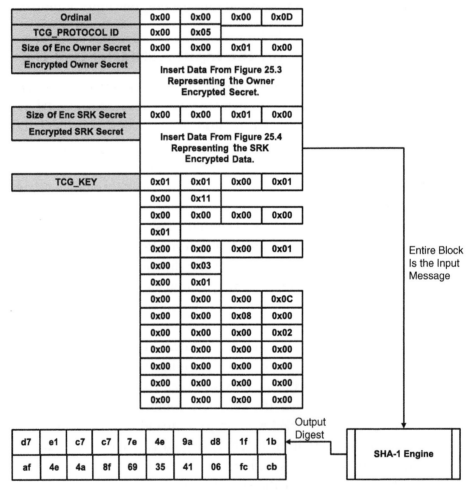

Figure 25.7 Calculating the payload hash for TPM_TakeOwnership

The next step is to build the input message that is to be used by the HMAC engine that will produce the final authorization digest for the TPM_TakeOwnership command. Figure 25.8 defines the compilation of

Parameter Digest	d7	e1	c7	c7	7e	4e	9a	d8	1f	1b
	af	4e	4a	8f	69	35	41	06	fc	cb
nonceEven	a5	a5	a5	a5	a5	a5	a5	a5	a5	a5
	a5	a5	a5	a5	a5	a5	a5	a5	a5	a5
nonceOdd	B9	73	05	FA	DB	E3	4D	C5	46	65
	10	00	0A	55	04	2E	3F	EA	BF	27
Continue Auth Session	01									

Figure 25.8   Input message to the HMAC engine

the input message for the HMAC engine using the payload digest, nonce Even, nonceOdd, and continue authorization session parameters.

Note that the Conformance Configuration Specification defines the two nonce values and the TPM will always produce the same nonceEven value as long as it is in compliance state. The only remaining step is to HMAC the input message with the Owner Secret. That secret is the HMAC key and is the same secret that was encrypted with the EK public key within the input message block. The TPM_TakeOwnership command is referred to as a deferred authorization command by virtue of the fact that prior to executing this command, the Owner has not been established and this command is Owner-authorized. The complete HMAC calculation is described in Figure 25.9.

This authorization digest should match the digest you calculate if the TPM is in compliance state and you use the predefined OAEP encoding along with the defined nonce values. If you do not get this 20-byte digest value, you might want to check your input parameters, SHA-1 digest calculation, and HMAC message, as well as with the HMAC key; the Owner Secret as defined by the Conformance Configuration Specification. The entire TPM_TakeOwnership command input message and authorization blocks are defined in Figure 25.10.

Step 4: Transmitting the Input Command Message

After you have the completed command input message block and authorization digest calculation, the next step is to transmit this information to the TPM. The vehicle used to accomplish this goal is the Trusted Device Driver

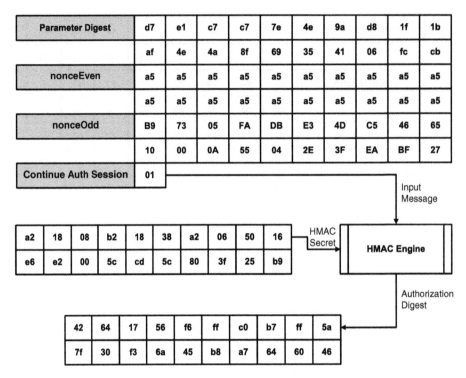

Parameter Digest	d7	e1	c7	c7	7e	4e	9a	d8	1f	1b
	af	4e	4a	8f	69	35	41	06	fc	cb
nonceEven	a5	a5	a5	a5	a5	a5	a5	a5	a5	a5
	a5	a5	a5	a5	a5	a5	a5	a5	a5	a5
nonceOdd	B9	73	05	FA	DB	E3	4D	C5	46	65
	10	00	0A	55	04	2E	3F	EA	BF	27
Continue Auth Session	01									

Figure 25.9 Calculating TPM_TakeOwnership authorization digest

Layer (TDDL), which encompasses the device driver, system management bus (SMBus) for those embedded types, and the defined Application Programming Interface (API). This example assumes that the driver has not opened the pipe and will go through all of the steps needed to communicate with the TPM device. You must first open a communication pipe with the TPM device via the TDDL API as shown here.

```
Tddli_Open()
```

This driver API call will establish a communication pipe with the TPM. Remember that only one pipe can be established with the TPM device at any given time and if you already executed this command, reexecuting it will return an error condition. Now that communication with the TPM has been established, you are ready to transmit the TPM_TakeOwnership command.

The command transmit API is defined next and takes four arguments:

```
Tddli_TransmitData( unsigned char *pInputCommandMessage,
                    UINT32 inputCommandMessageLength,
                    unsigned char *pcommandResultsBuffer,
                    UINT32 *pOutputResultsLength );
```

Authorization Tag	0x00	0xC2			
Parameter Size	0x00	0x00	0x02	0x70	
Ordinal	0x00	0x00	0x00	0x0D	
TCG_PROTOCOL ID	0x00	0x05			
Size of Enc Owner Secret	0x00	0x00	0x01	0x00	
Encrypted Owner Secret	Insert Encrypted Owner Secret Data from Figure 25.3.				
Size of Enc SRK Secret	0x00	0x00	0x01	0x00	
Encrypted SRK Secret	Insert Encrypted SRK Secret Data From Figure 25.4.				
TCG_KEY	0x01	0x01	0x00	0x01	
	0x00	0x11			
	0x00	0x00	0x00	0x00	
	0x01				
	0x00	0x00	0x00	0x01	
	0x00	0x03			
	0x00	0x01			
	0x00	0x00	0x00	0x0C	
	0x00	0x00	0x08	0x00	
	0x00	0x00	0x00	0x02	
	0x00	0x00	0x00	0x00	
	0x00	0x00	0x00	0x00	
	0x00	0x00	0x00	0x00	
	0x00	0x00	0x00	0x00	
Authorization Handle	0x00	0x00	0x00	0x00	
nonceOdd	0xB9	0x73	0x05	0xFA	0xDB
	0xE3	0x4D	0xC5	0x46	0x65
	0x10	0x00	0x0A	0x55	0x04
	0x2E	0x3F	0xEA	0xBF	0x27
Continue Auth Session	0x01				
Authorization Digest	0x42	0x64	0x17	0x56	0xf6
	0xff	0xc0	0xb7	0xff	0x5a
	0x7f	0x30	0xf3	0x6a	0x45
	0xb8	0xa7	0x64	0x60	0x46

Figure 25.10 Completed TPM_TakeOwnership input message block

This API simply points to a buffer that contains the TPM command input message and defines the command length. In addition, the API needs to know about the buffer that the TPM output message is to be written to and the size of this buffer on input. The resulting length of the output message will be populated on output response. See Chapter 22 for more detail concerning the TDDL API definition and usage.

When executing this command, TPM_TakeOwnership, note that the execution will generate an RSA key pair that will represent the SRK. This key generation will be instantaneous when in the compliance state, but when in the "normal" state, this command will have a more noticeable execution time because the TPM is generating this key pair. On a successful command execution, the API will return a TDDL_SUCCESS response and the command output message will be populated within the pointer to the **commandResultsBuffer**. Figure 25.11 defines the compliance command output message, including the attached authorization block.

Step 5: Verifying the Output Authorization Digest

The only remaining task that needs to be accomplished is the output authorization. This must be done by the TCG Software Stack (TSS), or embedded application, in order to trust the data that is being presented.

The output payload parameters must be used to calculate a SHA-1 payload digest, similar to the input payload digest calculation. Figure 25.12 defines the parameters that are to be used to define the input message block that the SHA-1 engine is to use to calculate the output payload digest. This digest will be used in a similar fashion relative to the input authorization digest calculation as described in previous pages in this chapter.

The next step is to produce the input message block for the HMAC engine. This input message block uses the newly calculated output payload digest, the new TPM nonceEven, the nonceOdd supplied during command input, and the continue authorization session supplied during command output. One note with regard to the nonceEven compliance value concerns the consistent 0xA5 numerical representation. This is per conformance definition, and if your TPM is in compliance state, you will use this nonceEven value as a result of every successful authorization command output execution. If your TPM is not in the compliance state, you will get back a truly random 20-byte value in regard to this parameter. The output digest calculation uses the same

Authorization Tag	0x00	0xC5														
Parameter Size	0x00	0x00	0x01	0x62												
Return Code	0x00	0x00	0x00	0x00												
TCG_KEY	0x01	0x01	0x00	0x02												
	0x00	0x11														
	0x00	0x00	0x00	0x00												
	0x01															
	0x00	0x00	0x00	0x01												
	0x00	0x03														
	0x00	0x01														
	0x00	0x00	0x00	0x0C												
	0x00	0x00	0x08	0x00												
	0x00	0x00	0x00	0x02												
	0x00	0x00	0x00	0x00												
	0x00	0x00	0x00	0x00												
	0x00	0x00	0x01	0x00												
	e2	f1	d9	e8	77	f6	f5	7f	0b	d0	08	9e	ba	37	37	c8
	31	01	d1	0d	20	b7	98	dd	26	91	f1	a1	5a	b5	31	c7
	11	86	71	95	f9	45	79	27	5a	5a	fb	a1	1c	3b	11	5f
	07	8a	59	53	e8	b6	67	bd	84	1d	9c	f1	e5	cd	71	51
	dd	9b	67	a7	d5	8d	3b	8a	e9	16	df	93	92	1f	7d	be
	d9	ab	f8	79	20	2a	29	0e	7d	f6	5b	71	d5	b2	6c	94
	6b	1e	fc	09	66	4f	8b	7c	0d	68	32	0e	e9	e7	cc	a0
	68	d0	c1	7e	4a	af	52	a5	4e	9e	16	34	1a	1a	6c	44
	40	8e	ec	67	fc	d5	49	1e	c7	78	63	a2	68	11	e2	3e
	e1	12	6b	80	9f	af	88	88	fd	5f	66	fd	12	68	b4	1f
	67	ec	15	6e	b1	a4	2e	29	40	dc	5a	d8	ab	a1	bb	5f
	75	28	69	8f	03	e2	b7	7f	44	70	3a	bc	6d	75	f4	10
	c8	e8	75	48	df	84	e8	0f	46	a9	1b	3d	8f	98	68	64
	48	b1	98	7e	85	1c	84	e1	92	44	ae	7d	79	ab	88	1e
	79	e9	88	c7	32	51	37	74	a8	99	ee	69	fb	bf	61	fd
	0a	b5	82	c6	19	53	36	44	9a	f9	a1	8f	3a	ca	db	9d
	0x00	0x00	0x00	0x00												
nonceEven	0xA5	0xA5	0xA5	0xA5	0xA5											
	0xA5	0xA5	0xA5	0xA5	0xA5											
	0xA5	0xA5	0xA5	0xA5	0xA5											
	0xA5	0xA5	0xA5	0xA5	0xA5											
Continue Auth Session	0x01															
Authorization Digest	0xf2	0x07	0x63	0xab	0x07											
	0x3b	0x10	0xdf	0x48	0x33											
	0x6a	0xc1	0xa8	0xa0	0x30											
	0xa5	0xa1	0xe1	0xaa	0x31											

Figure 25.11 TPM_TakeOwnership compliance output message block

Owner-authorization secret to determine the output authorization digest, as defined in Figure 25.13.

The last step is to compare the authorization digest returned from the TPM output message block to the authorization digest you calculated. If these digests match, you can continue with the TPM_TakeOwnership output message data. If these values do not match, first check your calculations; chances are that you made a mistake during the authorization digest calculation. The TPM_TakeOwnership output command message will return the public key representing the SRK that was generated within the TPM. Record this value since it will be used to load keying material into the TPM;

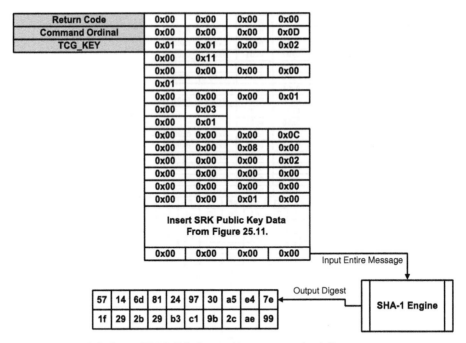

Figure 25.12 Calculating TPM_TakeOwnership output payload digest

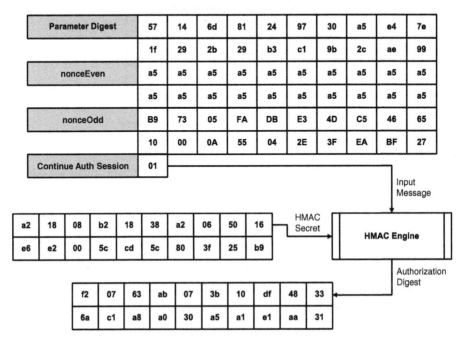

Figure 25.13 Calculating TPM_TakeOwnership output authorization digest

you can be guaranteed of using this public key once, since you must use the SRK as the top parent concerning your key management tree.

That's it; you have successfully taken ownership of the TPM device. This example has leveraged the TPM's compliance state to allow for predictable data, but this procedure is identical to what you would do during normal TPM command execution. One issue that you must take into account is the fact that the compliance state will produce consistent output data, and it would be very easy to mismanage these data items. The moral of the story is to make sure your data management utilities are doing what they are designed to do, especially with the nonceEven and nonceOdd values. If you mismanage these parameters, you will inevitably fail to authorize commands during subsequent execution of multiple command transmissions. The next and final chapter details some of the TPM commands used the most.

26

More Command Examples

This chapter expands on the previous one; it will leverage three compliance keys that are resident within the Trusted Platform Module (TPM) prior to execution of the TPM_ForceClear command: storage, signing, and binding key data, as well as the build TPM commands to leverage these entities. Note that the storage key will be used to generate an RSA key pair that can be loaded into the TPM. This type of command is designed for use in conjunction with the TPM_LoadKey command. The signing key will be used to generate a signature blob, and the binding key will be used to unbind a cryptographic blob.

Notice too that I provide the TPM_UnBind command – the cryptographic blob that is usually the result of successfully executing the TCG Software Stack (TSS) command TSS_Bind or using an external cryptographic engine – relative to the TPM. With that said, let's look at a command that leverages the use of the TPM storage key within the command TPM_CreateWrapKey.

26.1 TPM_CreateWrapKey Using Compliance Key 0

We are going to build and execute the TPM_CreateWrapKey command using the compliance key indexed by Key Handle 0x00000000, which is a storage key. The Conformance Configuration Specification defines the usage authorization for this key as Secret E, as shown in Figure 26.1.

This secret is used to determine the Hash-based Message Authentication Code (HMAC) key when creating the input and output authorization digests concerning the TPM_CreateWrapKey. See Chapter 15 for more details with regard to TPM_CreateWrapKey specifics. Note that the authorization session associated with this command must be of the type Object

Secret E									
0x89	0x99	0xa3	0xf5	0xa5	0x40	0xe1	0x45	0x86	0x4b
0x30	0x95	0x1b	0xc0	0x6f	0x8f	0xb0	0x60	0x01	0x69

Figure 26.1 Compliance Key Handle 0 usage authorization

Specific Authorization Protocol (OSAP); and before you can successfully execute the TPM_CreateWrapKey, you must create a handle to this session type. The OSAP will calculate a Shared Secret that will eventually be used as the HMAC key when authorizing the TPM message blocks. In addition, the Shared Secret is involved in the TCG_ENCAUTH data calculations that will encrypt the 20-byte usage and migration Authorization Secrets associated with the key about to be created. Figure 26.2 defines the TPM_OSAP command execution and Figure 26.3 defines the Shared Secret and encryption value that will be XORed with the Usage and Migration Secrets.

The next step is to "encrypt" the usage and migration of 20-byte secret values using the encryption value and an XOR function. This will protect the secret(s) value(s) during the command input transmission to the TPM. Figure 26.4 defines the Usage and Migration Secrets and the resulting encrypted values. These encrypted 20-byte values will be inserted into the input message block to make up a portion of the TPM_CreateWrapKey input payload.

The final payload parameter involves the type of keying material that we want the TPM to create. In this example we are going to command the TPM to create a signing key and the TCG_KEY structure will define this and other associated parameters involved in this key generation. Figure 26.5 describes the TCG_KEY structure population and Chapter 15 defines these parameters in detail.

This concludes the input command payload parameter population. The next task is to create an authorization block and digest that will allow this command to be authorized and executed within the TPM device.

The first step in attaining this goal is to create a contiguous input message that will be supplied to the Secure Hash Algorithm (SHA-1) engine and

Authorization Tag	0x00	0xC1			
Parameter Size	0x00	0x00	0x00	0x24	
Ordinal	0x00	0x00	0x00	0x0B	
Entity Type	0x00	0x05			
Entity Value	0x00	0x00	0x00	0x00	
nonceOddOSAP	0xB9	0x73	0x05	0xFA	0xDB
	0xE3	0x4D	0xC5	0x46	0x65
	0x10	0x00	0x0A	0x55	0x04
	0x2E	0x3F	0xEA	0xBF	0x27

Authorization Tag	0x00	0xC4			
Parameter Size	0x00	0x00	0x00	0x36	
Return Code	0x00	0x00	0x00	0x00	
Authorization Handle	0x00	0x00	0x00	0x00	
nonceEven	0xA5	0xA5	0xA5	0xA5	0xA5
	0xA5	0xA5	0xA5	0xA5	0xA5
	0xA5	0xA5	0xA5	0xA5	0xA5
	0xA5	0xA5	0xA5	0xA5	0xA5
nonceEvenOSAP	0xA5	0xA5	0xA5	0xA5	0xA5
	0xA5	0xA5	0xA5	0xA5	0xA5
	0xA5	0xA5	0xA5	0xA5	0xA5
	0xA5	0xA5	0xA5	0xA5	0xA5

Figure 26.2 TPM_OSAP execution

defined by the command input payload parameters. Figure 26.6 shows the message that will be supplied to the SHA-1 engine, which will produce the input payload digest.

Notice that the parent Key Handle is not part of the payload digest calculation, but this parameter must be supplied with the input command

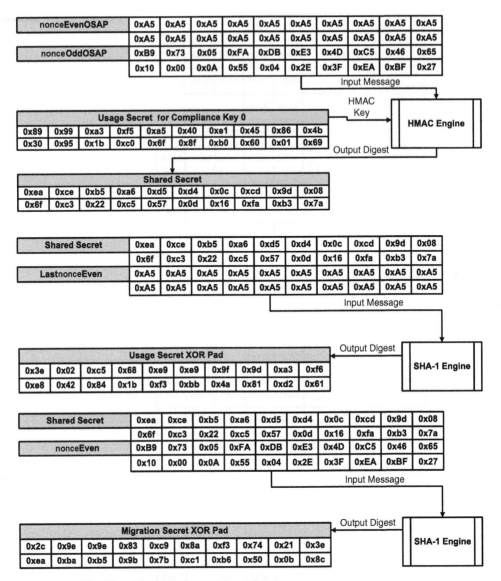

Figure 26.3 The Shared Secret and encryption values

message block. For embedded systems, storing this parameter and doing the payload digest calculation in place can save valuable resources. In other words, saving off Key Handle information will consume 4 bytes of data storage; the maximum data storage would be 8 bytes for two Key Handles, as opposed to the entire size of the command payload data. The next step is to

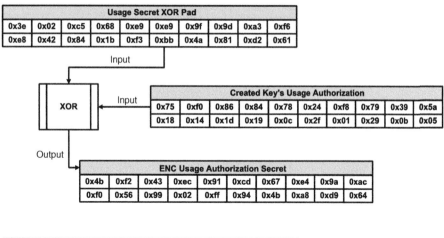

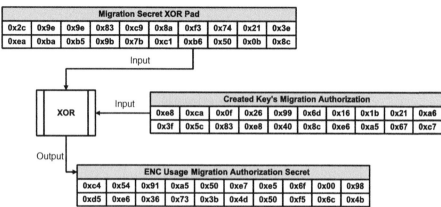

Figure 26.4 Handling the Usage and Migration Authorization Secrets

compose the authorization block and calculate the input command authorization digest.

We build the authorization block in the same manner as we built the TPM_TakeOwnership authorization area. The authorization block has the command payload digest, nonceEven, nonceOdd, continue authorization session, and a placeholder for the authorization digest. One note: this command will not allow the authorization session to be continued. Therefore, if you insert logic one (True) into this parameter, the output message block will declare the continuation of the authorization session to be False. This behavior is per-specification; any authorization sessions will be terminated on either a passing or failing command execution condition. If you want to create another RSA key pair using this command after executing a previous TPM_CreateWrapKey, you must reestablish the OSAP authorization

TCG_KEY Structure	0x01	0x01	0x01	0x03
	0x00	0x10		
	0x00	0x00	0x00	0x02
	0x01			
	0x00	0x00	0x00	0x01
	0x00	0x01		
	0x00	0x02		
	0x00	0x00	0x00	0x0C
	0x00	0x00	0x80	0x00
	0x00	0x00	0x00	0x02
	0x00	0x00	0x00	0x00
	0x00	0x00	0x00	0x00
	0x00	0x00	0x00	0x00
	0x00	0x00	0x00	0x00

Figure 26.5 TCG_KEY structure used in TPM_CreateWrapKey example

session. Figure 26.7 defines the authorization block parameters and digest calculation concerning this command.

Now that the authorization block has been created, simply build the entire TPM_CreateWrapKey input message block and send this data to the TPM. Refer to the TPM_TakeOwnership example regarding the Trusted Device Driver Layer (TDDL) Application Programming Interface (API) used to open a communication pipe and transmit the command data to the TPM. Again, the return response with regard to the successful execution of this command will be quick when the TPM is in the compliance state. If the TPM is in "normal" state, command completion will vary depending on the RSA engine in regard to key generation. Now let's look at the command output response concerning the TPM_CreateWrapKey execution results.

The output payload is simply a TCG_KEY structure that represents the public and wrapped private key portions of the RSA key generated by invoking the TPM_CreateWrapKey command. This wrapped key or TCG_KEY structure can be used directly by the TPM_LoadKey command input

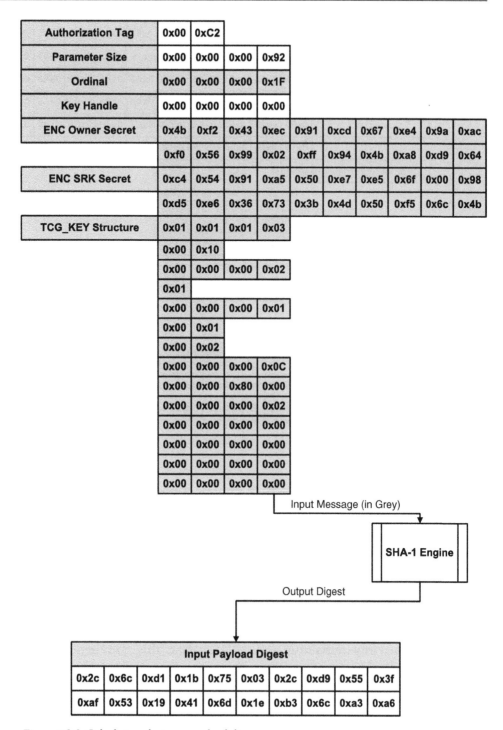

Figure 26.6 Calculating the input payload digest

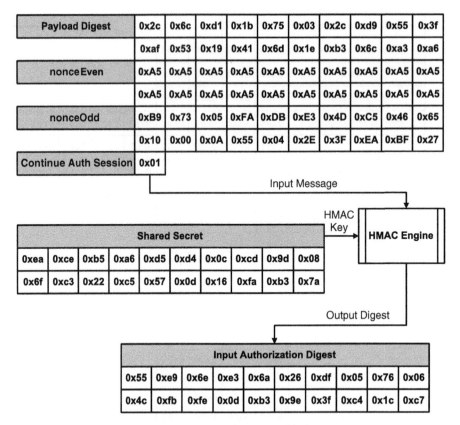

Payload Digest	0x2c	0x6c	0xd1	0x1b	0x75	0x03	0x2c	0xd9	0x55	0x3f
	0xaf	0x53	0x19	0x41	0x6d	0x1e	0xb3	0x6c	0xa3	0xa6
nonceEven	0xA5	0xA5	0xA5	0xA5	0xA5	0xA5	0xA5	0xA5	0xA5	0xA5
	0xA5	0xA5	0xA5	0xA5	0xA5	0xA5	0xA5	0xA5	0xA5	0xA5
nonceOdd	0xB9	0x73	0x05	0xFA	0xDB	0xE3	0x4D	0xC5	0x46	0x65
	0x10	0x00	0x0A	0x55	0x04	0x2E	0x3F	0xEA	0xBF	0x27
Continue Auth Session	0x01									

Input Message

HMAC Key → HMAC Engine

Shared Secret									
0xea	0xce	0xb5	0xa6	0xd5	0xd4	0x0c	0xcd	0x9d	0x08
0x6f	0xc3	0x22	0xc5	0x57	0x0d	0x16	0xfa	0xb3	0x7a

Output Digest

Input Authorization Digest									
0x55	0xe9	0x6e	0xe3	0x6a	0x26	0xdf	0x05	0x76	0x06
0x4c	0xfb	0xfe	0x0d	0xb3	0x9e	0x3f	0xc4	0x1c	0xc7

Figure 26.7 TPM_CreateWrapKey input authorization block

message block; I leave it as an exercise for readers to invoke this command. The output message block for the successful command execution in regard to the TPM_CreateWrapKey is defined in Figure 26.8.

Remember that the OSAP authorization session is terminated by the execution of the TPM_CreateWrapKey regardless of the state of the parameter **contAuthSession** contained in the input command message block. In addition, by virtue of the authorization specification, any failing command that requires an authorization session will terminate the session.

The only remaining task to complete this command execution is to authorize the output command message from the point of view of the TSS. Figure 26.9 defines the payload parameters that will be SHA-1 hashed to produce the output message payload digest.

Once we have the payload digest, we compile the input message block that will be input into the HMAC engine in an effort to calculate the output

```
00 c5
00 00 02 62
00 00 00 00
01 01 00 02
00 10
00 00 00 02
01
00 00 00 01
00 01
00 02
00 00 00 0c
00 00 08 00
00 00 00 02
00 00 00 00
00 00 00 00
00 00 01 00
ab 56 7c 0e 60 8c 5c 18 9e 90 2c 37 32 cf e3 fe
4f a7 b5 0c 78 a1 5d a7 39 eb c0 06 87 05 db 1f
e4 ab 2a 9a 68 e3 5b b6 fb 27 69 5a 4b e2 90 65
04 b2 78 cf 44 02 7c 16 4c fb f5 f0 f6 25 7d 31
f1 2e d8 67 93 5a 48 b2 c1 4c 16 fd 97 e5 86 65
4a 2e 07 4b 14 78 f7 66 83 66 05 b0 ea ec 1e 16
cf f9 f9 c5 5c bc 7b 42 24 a1 a7 1b 55 d7 4b b1
62 7f 90 88 ee fb fb 26 b1 4f 56 97 8c d0 12 05
a6 ef 09 c9 08 10 f2 1b 65 9c f2 05 7b cc 4e 6a
65 0c 1c e1 b5 3e 86 7d f8 0b 8b 6f e3 72 2b cb
c9 3d f8 61 f4 83 74 b1 38 a6 ce de 18 7f 8d c4
8f a1 8e a6 ac 71 a4 89 60 d3 3e 5f 3d 18 5c 32
6c 96 1d 84 8b 50 c3 5b 68 5c 16 2d 9c bb f1 79
60 6e c9 25 aa ec 26 9e 9e d4 d6 89 f3 ff 23 aa
75 46 3b 4a ea 1d e5 03 b9 ac 6d f8 2d 88 ff 84
12 b8 47 cf 3a 32 c9 66 c6 e3 2c 1f 7d 30 d8 99
00 00 01 00
91 1d c3 d4 99 48 55 cf cb a4 2a 01 bd 46 e9 ff
63 95 37 34 7b dc 5d cb 97 95 0e 05 24 a3 9a 52
ee 87 ae b0 27 ac 4d 64 3e 77 81 85 75 be 75 ae
ac 5d 6a 8d 04 41 23 bf 7c 45 83 3a e2 ba 06 b7
00 94 d8 af eb c8 a4 4a 55 60 24 7e 8b 61 9f c1
ba 37 b3 56 a0 b0 dd ad 81 e7 05 d6 34 1c 9a ec
a7 c2 09 84 0b a3 b4 a6 63 18 40 b7 7b 6e db 80
7c 8d fa a2 a4 7a de d7 57 ac cc 40 17 e3 74 c1
8f d9 fc e3 65 6b 98 39 35 7b 31 82 73 cb ed fc
37 19 ff 31 01 a3 0f b0 6b 1a 0f 5a c2 d6 6c af
20 cf 20 2c 1b c6 7d 0e fa 62 64 57 16 d8 92 59
00 05 93 1c 91 f3 69 b4 f2 29 44 9b b2 aa c3 ab
86 51 9b 8b 37 a2 4b 9c 26 58 19 64 d3 f3 e6 38
0d 9f c5 32 60 7a de 15 ba a9 49 87 7a 1a f4 b4
14 0c 22 ed 7a a5 30 c8 23 97 33 aa 5e 47 5a 9d
56 be 25 5c 35 2f f6 a3 0f c5 57 30 23 2a 84 5c
a5 a5 a5 a5 a5 a5 a5 a5 a5 a5 a5 a5 a5 a5 a5 a5 a5
00
a7 a8 f2 e2 37 66 3f 0c cf c8 85 d9 32 6c a2 e2 d4 93 c4 c3
```

Figure 26.8 TPM_CreateWrapKey output message block

command authorization digest. Note that the OSAP Shared Secret will be used as the HMAC key, since the authorization session attached to this command is an OSAP. Figure 26.10 defines the output authorization digest calculation using the payload digest, nonceEven, nonceOdd, and the continue

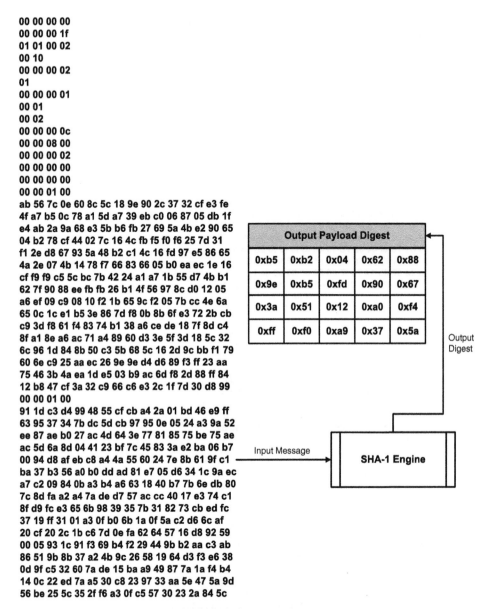

```
00 00 00 00
00 00 00 1f
01 01 00 02
00 10
00 00 00 02
01
00 00 00 01
00 01
00 02
00 00 00 0c
00 00 08 00
00 00 00 02
00 00 00 00
00 00 00 00
00 00 01 00
ab 56 7c 0e 60 8c 5c 18 9e 90 2c 37 32 cf e3 fe
4f a7 b5 0c 78 a1 5d a7 39 eb c0 06 87 05 db 1f
e4 ab 2a 9a 68 e3 5b b6 fb 27 69 5a 4b e2 90 65
04 b2 78 cf 44 02 7c 16 4c fb f5 f0 f6 25 7d 31
f1 2e d8 67 93 5a 48 b2 c1 4c 16 fd 97 e5 86 65
4a 2e 07 4b 14 78 f7 66 83 66 05 b0 ea ec 1e 16
cf f9 f9 c5 5c bc 7b 42 24 a1 a7 1b 55 d7 4b b1
62 7f 90 88 ee fb fb 26 b1 4f 56 97 8c d0 12 05
a6 ef 09 c9 08 10 f2 1b 65 9c f2 05 7b cc 4e 6a
65 0c 1c e1 b5 3e 86 7d f8 0b 8b 6f e3 72 2b cb
c9 3d f8 61 f4 83 74 b1 38 a6 ce de 18 7f 8d c4
8f a1 8e a6 ac 71 a4 89 60 d3 3e 5f 3d 18 5c 32
6c 96 1d 84 8b 50 c3 5b 68 5c 16 2d 9c bb f1 79
60 6e c9 25 aa ec 26 9e 9e d4 d6 89 f3 ff 23 aa
75 46 3b 4a ea 1d e5 03 b9 ac 6d f8 2d 88 ff 84
12 b8 47 cf 3a 32 c9 66 c6 e3 2c 1f 7d 30 d8 99
00 00 01 00
91 1d c3 d4 99 48 55 cf cb a4 2a 01 bd 46 e9 ff
63 95 37 34 7b dc 5d cb 97 95 0e 05 24 a3 9a 52
ee 87 ae b0 27 ac 4d 64 3e 77 81 85 75 be 75 ae
ac 5d 6a 8d 04 41 23 bf 7c 45 83 3a e2 ba 06 b7
00 94 d8 af eb c8 a4 4a 55 60 24 7e 8b 61 9f c1
ba 37 b3 56 a0 b0 dd ad 81 e7 05 d6 34 1c 9a ec
a7 c2 09 84 0b a3 b4 a6 63 18 40 b7 7b 6e db 80
7c 8d fa a2 a4 7a de d7 57 ac cc 40 17 e3 74 c1
8f d9 fc e3 65 6b 98 39 35 7b 31 82 73 cb ed fc
37 19 ff 31 01 a3 0f b0 6b 1a 0f 5a c2 d6 6c af
20 cf 20 2c 1b c6 7d 0e fa 62 64 57 16 d8 92 59
00 05 93 1c 91 f3 69 b4 f2 29 44 9b b2 aa c3 ab
86 51 9b 8b 37 a2 4b 9c 26 58 19 64 d3 f3 e6 38
0d 9f c5 32 60 7a de 15 ba a9 49 87 7a 1a f4 b4
14 0c 22 ed 7a a5 30 c8 23 97 33 aa 5e 47 5a 9d
56 be 25 5c 35 2f f6 a3 0f c5 57 30 23 2a 84 5c
```

Figure 26.9 Calculating the output payload digest for TPM_CreateWrapKey

authorization session. Finally, all that is left to do is compare the calculated authorization digest with the digest that was produced in the command output message block.

Again, the output of the TPM_CreateWrapKey, specifically the TCG_ KEY structure, can be used directly by the TPM_LoadKey command input

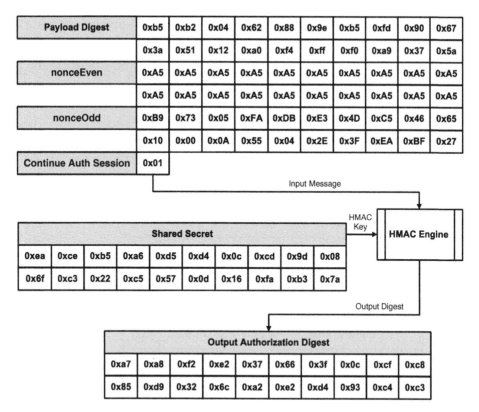

Figure 26.10 Calculating the output authorization digest for TPM_CreateWrapKey

message block. You can execute the TPM_LoadKey within the TPM compliance state and the resulting Key Handle will be the next available one after the defined compliance Key Handles.

26.2 TPM_Sign Using Compliance Key 2

The next example details an RSA signature via the successful execution of the TPM_Sign command. In this example, we are to sign a SHA-1 digest as opposed to signing the message directly, which would invoke a distinguished encoding role (DER) scheme. The idea is to SHA-1 the message and sign the digest attesting to the validity of its content. This method is not dependent on the length of the message to sign as opposed to the DER encoding of a signature to the message directly, which limits the size of the message that can be signed. The TPM_Sign input payload consists of two parameters: the size of the message to sign and the message itself.

The first task is to create an authorization session and resulting authorization handle that will be of the type Object Independent Authorization Protocol (OIAP). This authorization handle will be referred to during the command input and will allow the TPM to authorize the execution of the TPM_Sign

TPM_OIAP Input Message				
Authorization Tag	0x00	0xC1		
Parameter Size	0x00	0x00	0x00	0x0A
Ordinal	0x00	0x00	0x00	0x0A

TPM_OIAP Output Message					
Authorization Tag	0x00				
Parameter Size	0x00	0x00	0x00	0x22	
Return Code	0x00	0x00	0x00	0x00	
Authorization Handle	0x00	0x00	0x00	0x00	
nonceEven	0xA5	0xA5	0xA5	0xA5	0xA5
	0xA5	0xA5	0xA5	0xA5	0xA5
	0xA5	0xA5	0xA5	0xA5	0xA5
	0xA5	0xA5	0xA5	0xA5	0xA5

Figure 26.11 Executing TPM_OIAP

command. Figure 26.11 depicts the input and output message blocks concerning the execution of the TPM_OIAP command.

Note that in this example, the authorization handle is a 0x00000000 value and if you have another authorization session established prior to executing the TPM_OIAP command, you will not get this identifier concerning the handle. The point is to line up the authorization handle and the command that will refer this handle. You don't want the OAIP authorization handle to refer 0x00000000 and the TPM_Sign command to refer 0x00000001 – that would be bad.

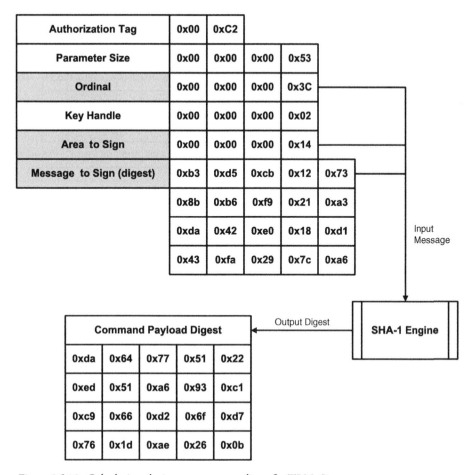

Figure 26.12 Calculating the input parameter digest for TPM_Sign

Now it is time to calculate the payload or input parameter digest and, as before, we must remove the parent Key Handle and establish a contiguous input message for our SHA-1 engine. Figure 26.12 defines the parameters that make up the input message that will be sent to our SHA-1 engine relative to the TPM_Sign input message block. Again, this 20-byte digest will be the first parameter in the input message block used by the HMAC engine in calculating the command authorization digest.

After we have the input command parameter digest, we can build the input message for the HMAC engine by concatenating the parameter digest, nonce-Even, nonceOdd, and the continue authorization session data parameters. The resulting HMAC input message block, Usage Secret, and resulting digest are defined in Figure 26.13.

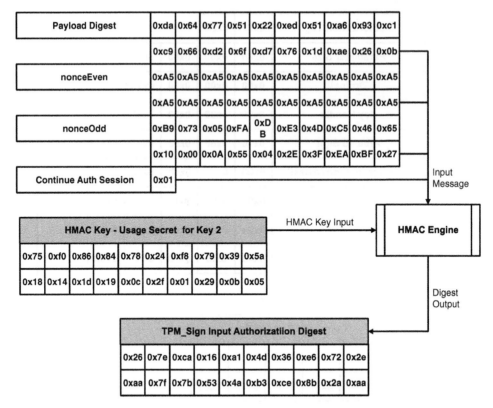

Figure 26.13 Calculating the TPM_Sign input authorization digest

Now that we have all of the parameters defined or calculated, the final task is to transmit the input command message to the TPM for execution. Figure 26.14 shows the entire TPM_Sign command input message that is to be sent to the TPM. The transmission mechanism is the same as in the previous example and the resulting output message is defined in Figure 26.15.

The command output message contains two payload parameters: the signature size and the resulting signature. Before we can use the data provided by the command output response, the authorization digest must be validated. As before, we must create a payload or parameter digest that will be used as an input parameter in regard to the HMAC calculation resulting in a calculated authorization digest. Figure 26.16 defines the parameters and the resulting SHA-1 payload digest values.

Now that we have the output response parameter digest, we can build the input message block that is to be fed to the HMAC engine. Figure 26.17 shows the HMAC input message block and the Usage Secret associated with

Authorization Tag	0x00	0xC2			
Parameter Size	0x00	0x00	0x00	0x53	
Ordinal	0x00	0x00	0x00	0x3C	
Key Handle	0x00	0x00	0x00	0x02	
Area to Sign	0x00	0x00	0x00	0x14	
Message to Sign (digest)	0xb3	0xd5	0xcb	0x12	0x73
	0x8b	0xb6	0xf9	0x21	0xa3
	0xda	0x42	0xe0	0x18	0xd1
	0x43	0xfa	0x29	0x7c	0xa6
Authorization Handle	0x00	0x00	0x00	0x02	
nonceOdd	0xB9	0x73	0x05	0xFA	0xDB
	0xE3	0x4D	0xC5	0x46	0x65
	0x10	0x00	0x0A	0x55	0x04
	0x2E	0x3F	0xEA	0xBF	0x27
Continue Auth Session	0x01				
Authorization Digest	0x26	0x7e	0xca	0x16	0xa1
	0x4d	0x36	0xe6	0x72	0x2e
	0xaa	0x7f	0x7b	0x53	0x4a
	0xb3	0xce	0x8b	0x2a	0xaa

Figure 26.14 TPM_Sign input message block

Key Handle 2 that results in the calculated value representing the authorization digest.

Once we have the calculated authorization digest, simply compare this value to the digest that was reported within the command output message block. If these authorization digests match, the signature can be consumed by the application requesting the signature operation.

Authorization Tag	0x00	0xC5									
Parameter Size	0x00	0x00	0x01	0x37							
Return Code	0x00	0x00	0x00	0x00							
Signature Size	0x00	0x00	0x01	0x00							
Signature	0xab	0xb3	0xd2	0xde	0x2e	0x3d	0xce	0xdd	0x9c	0xc6	0x34
	0x29	0x4c	0x52	0xbe	0x14	0xfb	0xf0	0x31	0xdd	0x22	0xa1
	0xb9	0xf7	0x9d	0xfe	0xe8	0x38	0x82	0xcd	0xff	0x37	0xfd
	0x57	0x35	0x70	0xee	0x63	0x42	0xa8	0xb4	0x2a	0xce	0xe7
	0x16	0x0a	0x84	0x1f	0xc6	0xf0	0x33	0x0b	0xca	0x80	0x61
	0x40	0xf5	0x52	0xf2	0xa6	0xbb	0xb5	0x93	0x35	0x91	0xbf
	0x8f	0xbc	0x48	0x2d	0x27	0x0b	0x8d	0x17	0x63	0xf6	0xa1
	0x86	0x9a	0xe5	0x27	0x57	0x09	0x02	0xc0	0x7f	0xb8	0xba
	0xc4	0x9a	0x5b	0xfe	0xdf	0xe3	0xb8	0xfd	0x7d	0x6d	0xb4
	0xd6	0x81	0x94	0xed	0xf6	0xe0	0x1e	0x54	0x30	0x0b	0x15
	0x79	0x70	0xb7	0xef	0x3d	0x15	0x94	0xd3	0xba	0xa6	0x99
	0xba	0xbf	0xae	0x6d	0x12	0xc0	0xc3	0xb3	0x13	0xf2	0x57
	0x21	0xc3	0x8e	0x3c	0x08	0x73	0x27	0xdd	0x87	0xb8	0x7a
	0xdd	0x9c	0xe2	0xc6	0x2b	0x15	0x7d	0x9a	0x87	0x31	0x8c
	0x2d	0x88	0xc8	0x58	0x06	0x8a	0x4d	0x7b	0x26	0xf7	0x4e
	0xd9	0x48	0x9f	0x5a	0x96	0x12	0x22	0x94	0x62	0xcd	0xb6
	0xa5	0x87	0x98	0xe1	0xb1	0x28	0x56	0xb3	0x21	0xb9	0xe5
	0x77	0xf3	0xc2	0x7a	0x3e	0xba	0x08	0xa0	0xb0	0x29	0x55
	0xa2	0x66	0x2a	0x5c	0xb2	0xb6	0x2f	0x36	0xa3	0xb5	0x6f
	0xb4	0x85	0xbe	0xa3	0xa3	0xe6	0x39	0x0c	0xb0	0x37	0x4b
	0x23	0x10	0x23	0xd2	0xff	0x81	0xf7	0x5a	0xeb	0x9b	0xb5
	0x70	0xde	0x97	0xbd	0x5a	0xf9	0x92	0xe6	0x33	0x81	0x17
	0x68	0x36	0xca	0xc5	0x60	0x6a	0x39	0x78	0x0c	0x53	0x45
	0xa3	0xf2	0xdd								
nonceEven	0xA5	0xA5	0xA5	0xA5	0xA5	0xA5	0xA5	0xA5	0xA5	0xA5	
	0xA5	0xA5	0xA5	0xA5	0xA5	0xA5	0xA5	0xA5	0xA5	0xA5	
Continue Auth Session	0x01										
Authorization Digest	0x5e	0x9d	0xd9	0x84	0x9c	0xd3	0xdb	0x01	0xce	0x57	
	0x79	0x97	0x24	0x7c	0x9c	0x8a	0x01	0x69	0x3c	0xb0	

Figure 26.15 TPM_Sign output message response

26.3 TPM_UnBind Using Compliance Key Handle 4

In this example, we are going to perform a sequence of commands. First is a TPM_LoadKey that will load in a binding key whose parent is the compliance key indexed by Key Handle 0x00000000. Next, we will unbind data that has been bound to the private key, indexed by the newly loaded key, using the associated public key. This example has the bound data that was encrypted

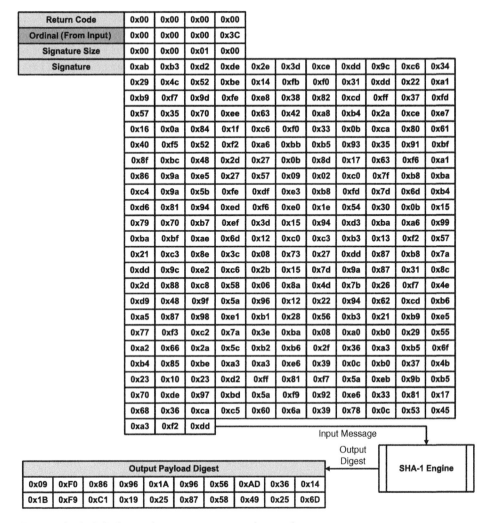

Figure 26.16 Calculating the output parameter digest value

by an external RSA cryptographic engine available to it; remember the TPM is a private key RSA engine. The first task is to load a binding key into the TPM. Note that this key is in addition to the compliance key indexed by Key Handle 0x00000004, which is a binding key as well.

The first order of business is to load the binding key into the TPM so that we can use this key to unbind RSA encrypted data. This involves the successful execution of the TPM_LoadKey that has an association with the public key that encrypted the data via the TSS_Bind or some other means.

Figure 26.18 shows the input command message block associated with the parameters that are to be hashed (shaded dark grey). Again, the first task

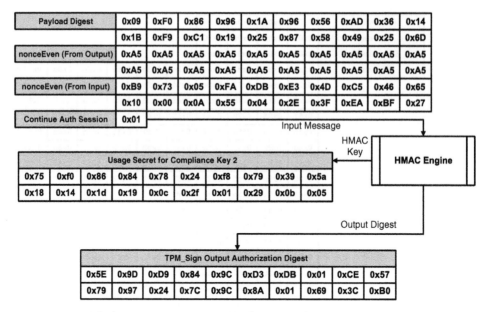

Figure 26.17 Calculating TPM_Sign output authorization digest

to complete is the hashing of the input command data parameters and Figure 26.19 defines the SHA-1 input message and the resulting digest.

Next, calculate the command input authorization digest by concatenating the parameter digest, nonceEven, nonceOdd, and the continue authorization session parameters. Figure 26.20 defines this authorization digest calculation using the previously described data and the Usage Secret associated with the parent or wrapping key.

Finally, we need to complete the input message concerning the TPM_LoadKey command and transmit this command to the TPM. Note that the TPM must have an OIAP session associated with this authorized command. So make sure that you execute the TPM_OIAP command and refer the resulting authorization handle within the input message of the TPM_LoadKey. The successful execution of the TPM_LoadKey command will result in the command output message shown in Figure 26.21.

Notice that the Key Handle is defined as 0x00000005, the next indexed location after the last defined compliance key; if you are executing within the compliance state, this is the handle to index when executing the TPM_UnBind command. Before executing this command, however, first validate the TPM_LoadKey output message. Figure 26.22 defines the expected

```
00 C2
00 00 02 6a
00 00 00 20
00 00 00 00
01 01 00 03
00 14
00 00 00 02
01
00 00 00 01
00 03
00 01
00 00 00 0c
00 00 08 00
00 00 00 02
00 00 00 00
00 00 00 00
00 00 01 00
bf 27 b1 3b d0 f5 a4 97 41 3c 7e 5d 9e 6a e4 73
c5 b7 7a 49 d3 a7 ff da f2 3b 6e 95 69 c0 8e 40
5f df d2 ce c3 b9 d4 6a 0e c1 4d 54 17 8d fd 63
38 ea 28 af 31 fc 22 2a 27 28 56 d1 6f d3 3b a0
46 05 65 9a 31 62 f1 14 ae 1f 3a 81 d0 31 a3 cf
ae f5 24 b4 a1 7c 9a 9b 3c ef 6c 98 3a fb 6f bf
58 03 05 58 49 63 a3 b0 7d d2 39 e8 15 f6 f3 40
ee 65 6b 10 7a fc 43 4c 24 96 32 63 37 d0 e5 c6
c5 05 43 b8 ca 35 30 22 61 a3 36 8d 29 35 a9 49
8e 82 6e 03 01 c3 99 93 8b 1b 07 dc b4 73 c4 9e
59 32 5b 2a bd 7b 54 21 4e 56 7a ca 70 a0 18 b1
f5 0f 57 81 d7 18 92 d7 73 31 4a 83 7f c1 0b 0a
6e a7 1d cf 53 48 ff 11 be 01 96 df 68 5b 63 1a
79 43 fe 3b 94 f4 db 5e f8 d8 e9 b0 f7 39 32 4e
c8 2e 0a 4f be 19 d8 6d 61 70 c0 a8 08 fc c2 4f
a5 86 03 6a 0f 67 51 9c 1b 49 7a 54 e7 68 1f eb
00 00 01 00
19 84 bb b1 27 29 69 65 46 2a 17 fb 4d 56 c0 4b
3b 01 75 c2 e6 9e 92 d5 86 8e d6 6f a8 99 97 7e
99 f9 b0 84 87 3c 28 aa 11 b9 cc 24 f6 3c a8 c9
15 79 11 94 ff c0 43 21 a0 1d 31 f2 ab 08 dc 98
a0 ea 55 01 ea 2b ad 9b 52 6b 25 19 a5 e1 5d dc
6e f6 f3 e6 b2 f9 1e 8f 28 f4 2a 71 47 f9 34 7f
a4 c8 c1 f6 18 a1 3a 2e a7 ad 15 7f df b0 45 5a
51 10 34 11 09 65 f1 21 eb aa 92 64 5c 17 e0 a6
b7 56 12 97 ec 9d 40 2e 94 a5 aa 4d 8d 31 fc f6
49 15 58 2f d1 49 46 81 c9 bd ef aa 83 fc 77 19
7b 75 78 c9 35 23 a3 26 93 47 4a bd ba 43 6f ea
33 26 e2 4e ac 44 26 22 73 20 80 1a d9 b0 e8 2e
16 a3 f9 7c 48 72 83 b2 94 f0 1b 8e c9 f9 16 71
d1 e6 75 42 80 05 e9 bd d5 d9 6c e2 52 72 d3 32
d0 3f 3e 20 e3 3b 40 a2 07 3a 58 68 71 e1 10 54
c0 e1 b6 b4 4c 71 c0 4f d4 e1 ec 1d 43 bb 5d 5b
00 00 00 00
B9 73 05 FA DB E3 4D C5 46 65 10 00 0A 55 04 2E 3F EA BF 27
01
38 ca 4b 5f 3d 57 cd d0 c5 1f 2e 14 29 ac ce aa 6f c8 36 7a
```

Figure 26.18 Example TPM_LoadKey input message block

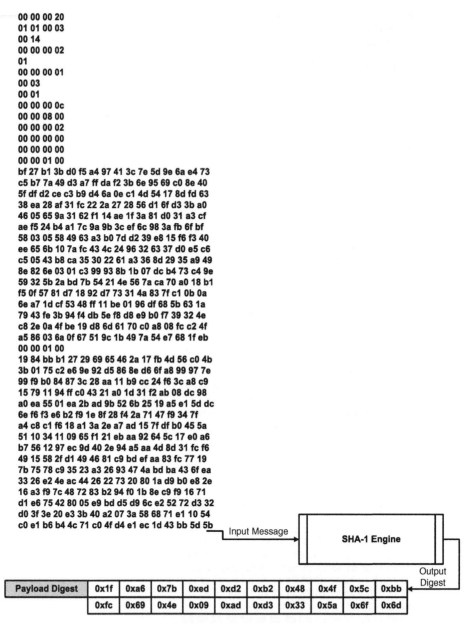

```
00 00 00 20
01 01 00 03
00 14
00 00 00 02
01
00 00 00 01
00 03
00 01
00 00 00 0c
00 00 08 00
00 00 00 02
00 00 00 00
00 00 00 00
00 00 01 00
bf 27 b1 3b d0 f5 a4 97 41 3c 7e 5d 9e 6a e4 73
c5 b7 7a 49 d3 a7 ff da f2 3b 6e 95 69 c0 8e 40
5f df d2 ce c3 b9 d4 6a 0e c1 4d 54 17 8d fd 63
38 ea 28 af 31 fc 22 2a 27 28 56 d1 6f d3 3b a0
46 05 65 9a 31 62 f1 14 ae 1f 3a 81 d0 31 a3 cf
ae f5 24 b4 a1 7c 9a 9b 3c ef 6c 98 3a fb 6f bf
58 03 05 58 49 63 a3 b0 7d d2 39 e8 15 f6 f3 40
ee 65 6b 10 7a fc 43 4c 24 96 32 63 37 d0 e5 c6
c5 05 43 b8 ca 35 30 22 61 a3 36 8d 29 35 a9 49
8e 82 6e 03 01 c3 99 93 8b 1b 07 dc b4 73 c4 9e
59 32 5b 2a bd 7b 54 21 4e 56 7a ca 70 a0 18 b1
f5 0f 57 81 d7 18 92 d7 73 31 4a 83 7f c1 0b 0a
6e a7 1d cf 53 48 ff 11 be 01 96 df 68 5b 63 1a
79 43 fe 3b 94 f4 db 5e f8 d8 e9 b0 f7 39 32 4e
c8 2e 0a 4f be 19 d8 6d 61 70 c0 a8 08 fc c2 4f
a5 86 03 6a 0f 67 51 9c 1b 49 7a 54 e7 68 1f eb
00 00 01 00
19 84 bb b1 27 29 69 65 46 2a 17 fb 4d 56 c0 4b
3b 01 75 c2 e6 9e 92 d5 86 8e d6 6f a8 99 97 7e
99 f9 b0 84 87 3c 28 aa 11 b9 cc 24 f6 3c a8 c9
15 79 11 94 ff c0 43 21 a0 1d 31 f2 ab 08 dc 98
a0 ea 55 01 ea 2b ad 9b 52 6b 25 19 a5 e1 5d dc
6e f6 f3 e6 b2 f9 1e 8f 28 f4 2a 71 47 f9 34 7f
a4 c8 c1 f6 18 a1 3a 2e a7 ad 15 7f df b0 45 5a
51 10 34 11 09 65 f1 21 eb aa 92 64 5c 17 e0 a6
b7 56 12 97 ec 9d 40 2e 94 a5 aa 4d 8d 31 fc f6
49 15 58 2f d1 49 46 81 c9 bd ef aa 83 fc 77 19
7b 75 78 c9 35 23 a3 26 93 47 4a bd ba 43 6f ea
33 26 e2 4e ac 44 26 22 73 20 80 1a d9 b0 e8 2e
16 a3 f9 7c 48 72 83 b2 94 f0 1b 8e c9 f9 16 71
d1 e6 75 42 80 05 e9 bd d5 d9 6c e2 52 72 d3 32
d0 3f 3e 20 e3 3b 40 a2 07 3a 58 68 71 e1 10 54
c0 e1 b6 b4 4c 71 c0 4f d4 e1 ec 1d 43 bb 5d 5b
```

Input Message → **SHA-1 Engine** → Output Digest

Payload Digest	0x1f	0xa6	0x7b	0xed	0xd2	0xb2	0x48	0x4f	0x5c	0xbb
	0xfc	0x69	0x4e	0x09	0xad	0xd3	0x33	0x5a	0x6f	0x6d

Figure 26.19 Calculating TPM_LoadKey parameter digest

TPM_LoadKey output message and the parameters that will be hashed to produce that parameter digest.

Next, build the input message for the HMAC engine using the parameter digest, nonceEven, nonceOdd, and continue authorization session

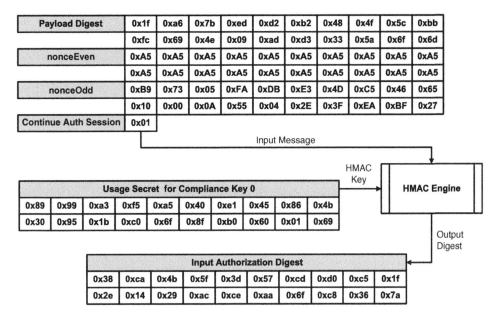

Payload Digest	0x1f	0xa6	0x7b	0xed	0xd2	0xb2	0x48	0x4f	0x5c	0xbb
	0xfc	0x69	0x4e	0x09	0xad	0xd3	0x33	0x5a	0x6f	0x6d
nonceEven	0xA5	0xA5	0xA5	0xA5	0xA5	0xA5	0xA5	0xA5	0xA5	0xA5
	0xA5	0xA5	0xA5	0xA5	0xA5	0xA5	0xA5	0xA5	0xA5	0xA5
nonceOdd	0xB9	0x73	0x05	0xFA	0xDB	0xE3	0x4D	0xC5	0x46	0x65
	0x10	0x00	0x0A	0x55	0x04	0x2E	0x3F	0xEA	0xBF	0x27
Continue Auth Session	0x01									

Input Message

HMAC Key

HMAC Engine

Usage Secret for Compliance Key 0									
0x89	0x99	0xa3	0xf5	0xa5	0x40	0xe1	0x45	0x86	0x4b
0x30	0x95	0x1b	0xc0	0x6f	0x8f	0xb0	0x60	0x01	0x69

Output Digest

Input Authorization Digest									
0x38	0xca	0x4b	0x5f	0x3d	0x57	0xcd	0xd0	0xc5	0x1f
0x2e	0x14	0x29	0xac	0xce	0xaa	0x6f	0xc8	0x36	0x7a

Figure 26.20   Calculating the TPM_LoadKey input authorization digest

Authorization Tag	0x00	0xC5			
Parameter Size	0x00	0x00	0x00	0x37	
Return Code	0x00	0x00	0x00	0x00	
Key Handle	0x00	0x00	0x00	0x05	
nonceEven	0xA5	0xA5	0xA5	0xA5	0xA5
	0xA5	0xA5	0xA5	0xA5	0xA5
	0xA5	0xA5	0xA5	0xA5	0xA5
	0xA5	0xA5	0xA5	0xA5	0xA5
Continue Auth Session	0x01				
Authorization Digest	0xd0	0xbc	0x61	0x8b	0x3b
	0x6d	0x97	0x2c	0x6e	0xe1
	0xf8	0x69	0xde	0xa5	0x1d
	0x87	0x8d	0x28	0xa3	0x17

Figure 26.21   The resulting TPM_LoadKey output response

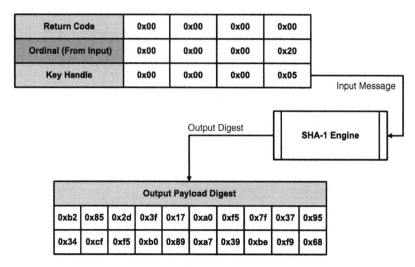

Return Code	0x00	0x00	0x00	0x00
Ordinal (From Input)	0x00	0x00	0x00	0x20
Key Handle	0x00	0x00	0x00	0x05

Input Message

Output Digest

SHA-1 Engine

Output Payload Digest

0xb2	0x85	0x2d	0x3f	0x17	0xa0	0xf5	0x7f	0x37	0x95
0x34	0xcf	0xf5	0xb0	0x89	0xa7	0x39	0xbe	0xf9	0x68

Figure 26.22 Calculating the output parameter digest of TPM_LoadKey

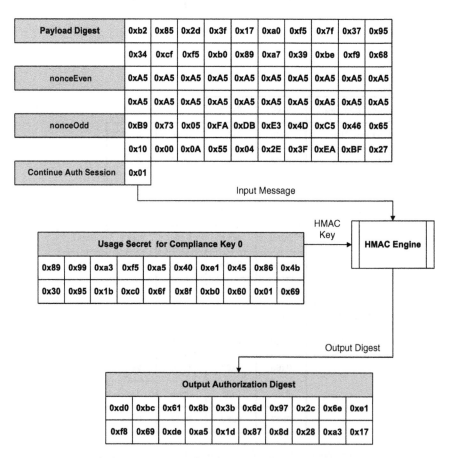

Payload Digest	0xb2	0x85	0x2d	0x3f	0x17	0xa0	0xf5	0x7f	0x37	0x95
	0x34	0xcf	0xf5	0xb0	0x89	0xa7	0x39	0xbe	0xf9	0x68
nonceEven	0xA5	0xA5	0xA5	0xA5	0xA5	0xA5	0xA5	0xA5	0xA5	0xA5
	0xA5	0xA5	0xA5	0xA5	0xA5	0xA5	0xA5	0xA5	0xA5	0xA5
nonceOdd	0xB9	0x73	0x05	0xFA	0xDB	0xE3	0x4D	0xC5	0x46	0x65
	0x10	0x00	0x0A	0x55	0x04	0x2E	0x3F	0xEA	0xBF	0x27
Continue Auth Session	0x01									

Input Message

Usage Secret for Compliance Key 0

0x89	0x99	0xa3	0xf5	0xa5	0x40	0xe1	0x45	0x86	0x4b
0x30	0x95	0x1b	0xc0	0x6f	0x8f	0xb0	0x60	0x01	0x69

HMAC Key

HMAC Engine

Output Digest

Output Authorization Digest

0xd0	0xbc	0x61	0x8b	0x3b	0x6d	0x97	0x2c	0x6e	0xe1
0xf8	0x69	0xde	0xa5	0x1d	0x87	0x8d	0x28	0xa3	0x17

Figure 26.23 Calculating TPM_LoadKey output authorization digest

Authorization Tag	0x00	0xC2									
Parameter Size	0x00	0x00	0x01	0x3f							
Ordinal	0x00	0x00	0x00	0x1e							
Key Handle	0x00	0x00	0x00	0x05							
Blob Length	0x00	0x00	0x01	0x00							
Blob to UnBind	0xa5	0x87	0x79	0x31	0x29	0x52	0x61	0x68	0x44	0xa7	0xc2
	0x3a	0xa1	0xce	0xf2	0x31	0xbe	0x4f	0x52	0xfe	0x1a	0x29
	0x62	0x63	0x00	0x24	0x96	0xac	0xdb	0x52	0xec	0x5a	0xdb
	0x0c	0x97	0xa9	0xce	0x35	0x29	0x3c	0x9e	0x0b	0xe8	0xe7
	0xa9	0x13	0x5e	0x31	0x01	0xdd	0x19	0x16	0x9f	0x33	0xab
	0x84	0x85	0x76	0x29	0xed	0x6d	0x7f	0x0b	0xa8	0xed	0x16
	0xef	0x17	0x0f	0xa0	0xcb	0x5f	0xac	0xd3	0x1a	0x32	0xc4
	0x88	0xcb	0x29	0x7d	0x60	0xce	0x2b	0x07	0x5a	0x23	0x1d
	0xc1	0x1d	0x09	0x42	0x3a	0x83	0xa3	0x62	0x60	0xd9	0x2d
	0xf9	0xbe	0x66	0xf1	0x49	0x3a	0x14	0x0b	0xc5	0x4c	0x20
	0x70	0x3b	0x3d	0xaf	0x20	0x2a	0x5d	0x79	0xd0	0xa4	0x71
	0x46	0x3f	0x14	0xd8	0x9b	0x40	0x5b	0x6f	0x3e	0x44	0xe3
	0x10	0xb9	0x60	0x8e	0x96	0x87	0x05	0xbe	0xc1	0x92	0x7d
	0x69	0x97	0x9b	0x92	0xc9	0xe4	0xb1	0x56	0x87	0x61	0x8e
	0xdd	0xe2	0xed	0x30	0xa4	0x62	0x32	0x50	0x73	0x4f	0xd5
	0xff	0xce	0x31	0x78	0x30	0x86	0x8d	0xae	0x78	0x08	0x08
	0x86	0xaa	0x3e	0xdd	0xb8	0x02	0x8a	0x2e	0x35	0xa2	0x0e
	0xde	0xc5	0x7a	0x7a	0x37	0xed	0x1b	0x8e	0x68	0x7b	0xe6
	0xb7	0xa8	0x8b	0x88	0x2d	0x4f	0x73	0xba	0x4e	0x55	0xbc
	0x2e	0xa0	0xc2	0xff	0x58	0xf9	0xaa	0x34	0x4a	0x23	0x98
	0x14	0x31	0xed	0xde	0x7b	0xd1	0x2b	0x4f	0x3b	0x63	0x38
	0x8e	0x33	0x93	0x95	0x82	0x90	0xfa	0x7c	0x61	0x52	0x69
	0xe1	0x63	0x73	0xd9	0x07	0x75	0xe3	0x43	0x7d	0x51	0x67
	0xb3	0xfc	0x57								

Figure 26.24 TPM_UnBind input message

parameters. Figure 26.23 defines the HMAC input message, Usage Secret and the resulting calculated authorization digest. One note: the authorization secret associated with the parent key, Key Handle 0x00000000, is defined by the Compliance Configuration Specification as Secret E.

Make sure that the calculated digest matches the authorization digest returned within the TPM_LoadKey command output message block; if so, we are ready to execute the TPM_UnBind command.

The TPM_UnBind command is very straightforward; the input command payload is defined as the encrypted blob size and the encrypted blob. Figure 26.24 defines the input message block concerning this command; notice that we are using Key Handle 0x00000005, which is the key we just

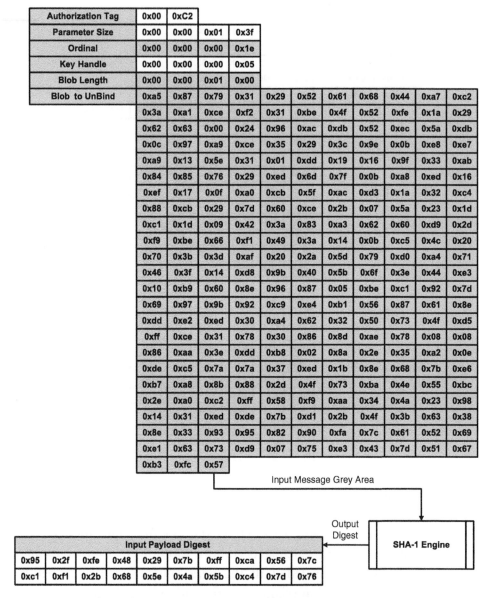

Figure 26.25 Calculated input parameter digest for TPM_UnBind input message

loaded into the TPM. In this example, note that I executed another TPM_ OIAP command that produced another OIAP session. This session is indexed by the authorization handle 0x00000001. You could use the same OIAP instead of generating another authorization session; remember that the OIAP session is object-independent. I chose this method – generating two authorization sessions – to keep the example simple and understandable.

The parameter digest must be calculated next. Figure 26.25 defines the SHA-1 input message and the resulting digest that represents the input command parameter digest.

Finally, we can calculate the input command authorization digest using the parameter digest, nonceEven, nonceOdd, and continue authorization session. Figure 26.26 defines this calculation. Note that the Usage Secret concerning the newly loaded key has not been defined; it is defined by the Compliance Configuration Specification as Secret G.

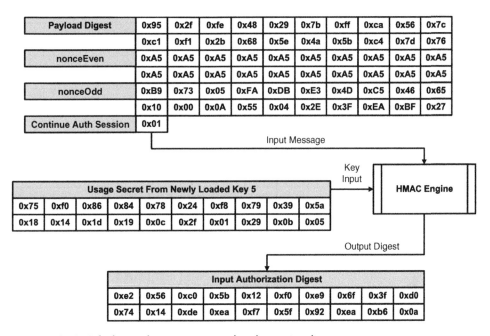

Figure 26.26 Calculating the input command authorization digest

Now all we have to do is to compile the complete TPM_UnBind command and send it to the TPM for authorization and execution. Figure 26.27 shows the entire TPM_UnBind input command message. When the TPM has completed the command, the response will be the RSA decrypted data, as defined in the figure.

The only issue we have left at this point is to validate the output authorization digest. This is accomplished in the same manner as in the other examples – calculate the parameters digest and concatenate this value with the nonceEven, nonceOdd, and continue authorization session parameters. Next, simply HMAC this input message and use the newly loaded Usage

Authorization Tag	0x00	0xC2									
Parameter Size	0x00	0x00	0x01	0x3f							
Ordinal	0x00	0x00	0x00	0x1e							
Key Handle	0x00	0x00	0x00	0x05							
Blob Length	0x00	0x00	0x01	0x00							
Blob to UnBind	0xa5	0x87	0x79	0x31	0x29	0x52	0x61	0x68	0x44	0xa7	0xc2
	0x3a	0xa1	0xce	0xf2	0x31	0xbe	0x4f	0x52	0xfe	0x1a	0x29
	0x62	0x63	0x00	0x24	0x96	0xac	0xdb	0x52	0xec	0x5a	0xdb
	0x0c	0x97	0xa9	0xce	0x35	0x29	0x3c	0x9e	0x0b	0xe8	0xe7
	0xa9	0x13	0x5e	0x31	0x01	0xdd	0x19	0x16	0x9f	0x33	0xab
	0x84	0x85	0x76	0x29	0xed	0x6d	0x7f	0x0b	0xa8	0xed	0x16
	0xef	0x17	0x0f	0xa0	0xcb	0x5f	0xac	0xd3	0x1a	0x32	0xc4
	0x88	0xcb	0x29	0x7d	0x60	0xce	0x2b	0x07	0x5a	0x23	0x1d
	0xc1	0x1d	0x09	0x42	0x3a	0x83	0xa3	0x62	0x60	0xd9	0x2d
	0xf9	0xbe	0x66	0xf1	0x49	0x3a	0x14	0x0b	0xc5	0x4c	0x20
	0x70	0x3b	0x3d	0xaf	0x20	0x2a	0x5d	0x79	0xd0	0xa4	0x71
	0x46	0x3f	0x14	0xd8	0x9b	0x40	0x5b	0x6f	0x3e	0x44	0xe3
	0x10	0xb9	0x60	0x8e	0x96	0x87	0x05	0xbe	0xc1	0x92	0x7d
	0x69	0x97	0x9b	0x92	0xc9	0xe4	0xb1	0x56	0x87	0x61	0x8e
	0xdd	0xe2	0xed	0x30	0xa4	0x62	0x32	0x50	0x73	0x4f	0xd5
	0xff	0xce	0x31	0x78	0x30	0x86	0x8d	0xae	0x78	0x08	0x08
	0x86	0xaa	0x3e	0xdd	0xb8	0x02	0x8a	0x2e	0x35	0xa2	0x0e
	0xde	0xc5	0x7a	0x7a	0x37	0xed	0x1b	0x8e	0x68	0x7b	0xe6
	0xb7	0xa8	0x8b	0x88	0x2d	0x4f	0x73	0xba	0x4e	0x55	0xbc
	0x2e	0xa0	0xc2	0xff	0x58	0xf9	0xaa	0x34	0x4a	0x23	0x98
	0x14	0x31	0xed	0xde	0x7b	0xd1	0x2b	0x4f	0x3b	0x63	0x38
	0x8e	0x33	0x93	0x95	0x82	0x90	0xfa	0x7c	0x61	0x52	0x69
	0xe1	0x63	0x73	0xd9	0x07	0x75	0xe3	0x43	0x7d	0x51	0x67
	0xb3	0xfc	0x57								
Authorization Handle	0x00	0x00	0x00	0x00							
nonceOdd	0xB9	0x73	0x05	0xFA	0xDB	0xE3	0x4D	0xC5	0x46	0x65	
	0x10	0x00	0x0A	0x55	0x04	0x2E	0x3F	0xEA	0xBF	0x27	
Continue Auth Session	0x01										
Authorization Digest	0xe2	0x56	0xc0	0x5b	0x12	0xf0	0xe9	0x6f	0x3f	0xd0	
	0x74	0x14	0xde	0xea	0xf7	0x5f	0x92	0xea	0xb6	0x0a	

Figure 26.27 Complete TPM_UnBind input message

Secret keys to complete the authorization calculation. Figure 26.28 shows this procedure.

As before, we must compare the calculated authorization digest with the digest supplied by the TPM_UnBind command output response. If the two digests match, we can use the decrypted data (see Figure 26.29).

Authorization Tag	0x00	0xC5									
Parameter Size	0x00	0x00	0x00	0xB8							
Return Code	0x00	0x00	0x00	0x00							
Decrypted Mess Length	0x00	0x00	0x00	0x81							
Message	0x54	0x68	0x65	0x20	0x6d	0x6f	0x73	0x74	0x20	0x76	0x61
	0x6c	0x75	0x61	0x62	0x6c	0x65	0x20	0x76	0x61	0x72	0x69
	0x65	0x74	0x79	0x20	0x6f	0x66	0x20	0x63	0x6f	0x72	0x75
	0x6e	0x64	0x75	0x6d	0x20	0x69	0x73	0x20	0x20	0x72	0x75
	0x62	0x79	0x2e	0x20	0x54	0x68	0x65	0x20	0x6e	0x61	0x6d
	0x65	0x20	0x63	0x6f	0x6d	0x65	0x73	0x20	0x66	0x72	0x6f
	0x6d	0x20	0x74	0x68	0x65	0x20	0x4c	0x61	0x74	0x69	0x6e
	0x20	0x72	0x75	0x62	0x72	0x75	0x6d	0x2c	0x20	0x22	0x72
	0x65	0x64	0x2e	0x22	0x20	0x49	0x74	0x20	0x68	0x61	0x73
	0x20	0x61	0x6c	0x73	0x6f	0x20	0x62	0x65	0x65	0x6e	0x20
	0x63	0x61	0x6c	0x6c	0x65	0x64	0x20	0x63	0x61	0x72	0x62
	0x75	0x6e	0x63	0x75	0x6c	0x75	0x73	0x2e			
nonceEven	0xA5	0xA5	0xA5	0xA5	0xA5	0xA5	0xA5	0xA5	0xA5	0xA5	
	0xA5	0xA5	0xA5	0xA5	0xA5	0xA5	0xA5	0xA5	0xA5	0xA5	
Continue Auth Session	0x01										
Authorization Digest	0xef	0xb3	0xa1	0x54	0x5e	0x6c	0x48	0xe6	0xca	0xd7	
	0xb9	0x9e	0x4a	0x4a	0x6e	0x8d	0xb8	0x60	0xa9	0x0a	

Figure 26.28 TPM_UnBind output response

Remember, all TPM commands have distinct rules; if you follow the given examples and apply the authorization techniques for authorized commands, you will be 60% complete concerning the TPM Command Suite. The other 40% involves understanding the specific command requirements and generating command input data that satisfies them.

Try expanding these examples; for instance, load in a signing key using the Storage Root Key (SRK) as a parent key or load another storage key; then with that key load in another signing key. The point is to exercise new knowledge to expand your understanding of the TPM and command compilation. Experience is king and the more commands you complete successfully, through trial and error, the more you will understand the TPM and its capabilities.

In addition, your TPM vendor has many years of experience both in designing the TPM device and in helping customers integrate it into their systems. I highly recommend that you establish a relationship with a TPM vendor and make a call if you find yourself in a situation that produces more questions than answers. Then too, you can email questions to me at my publisher's address. Most of all, don't let the TPM intimidate you; this device is

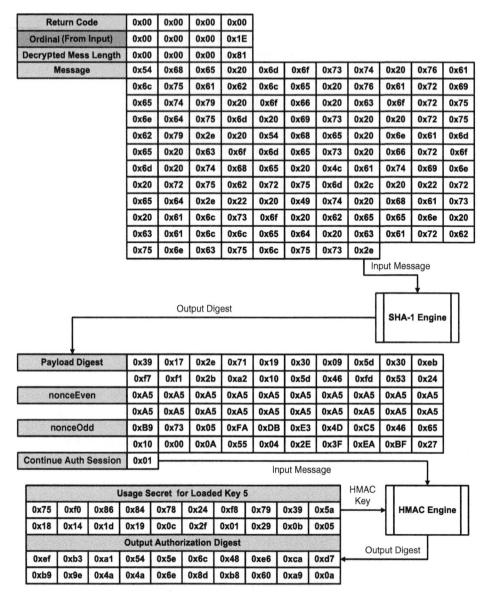

Return Code	0x00	0x00	0x00	0x00						
Ordinal (From Input)	0x00	0x00	0x00	0x1E						
Decrypted Mess Length	0x00	0x00	0x00	0x81						
Message	0x54	0x68	0x65	0x20	0x6d	0x6f	0x73	0x74	0x20	0x76	0x61
	0x6c	0x75	0x61	0x62	0x6c	0x65	0x20	0x76	0x61	0x72	0x69
	0x65	0x74	0x79	0x20	0x6f	0x66	0x20	0x63	0x6f	0x72	0x75
	0x6e	0x64	0x75	0x6d	0x20	0x69	0x73	0x20	0x20	0x72	0x75
	0x62	0x79	0x2e	0x20	0x54	0x68	0x65	0x20	0x6e	0x61	0x6d
	0x65	0x20	0x63	0x6f	0x6d	0x65	0x73	0x20	0x66	0x72	0x6f
	0x6d	0x20	0x74	0x68	0x65	0x20	0x4c	0x61	0x74	0x69	0x6e
	0x20	0x72	0x75	0x62	0x72	0x75	0x6d	0x2c	0x20	0x22	0x72
	0x65	0x64	0x2e	0x22	0x20	0x49	0x74	0x20	0x68	0x61	0x73
	0x20	0x61	0x6c	0x73	0x6f	0x20	0x62	0x65	0x65	0x6e	0x20
	0x63	0x61	0x6c	0x6c	0x65	0x64	0x20	0x63	0x61	0x72	0x62
	0x75	0x6e	0x63	0x75	0x6c	0x75	0x73	0x2e			

Input Message

SHA-1 Engine

Output Digest

Payload Digest	0x39	0x17	0x2e	0x71	0x19	0x30	0x09	0x5d	0x30	0xeb
	0xf7	0xf1	0x2b	0xa2	0x10	0x5d	0x46	0xfd	0x53	0x24
nonceEven	0xA5	0xA5	0xA5	0xA5	0xA5	0xA5	0xA5	0xA5	0xA5	0xA5
	0xA5	0xA5	0xA5	0xA5	0xA5	0xA5	0xA5	0xA5	0xA5	0xA5
nonceOdd	0xB9	0x73	0x05	0xFA	0xDB	0xE3	0x4D	0xC5	0x46	0x65
	0x10	0x00	0x0A	0x55	0x04	0x2E	0x3F	0xEA	0xBF	0x27
Continue Auth Session	0x01									

Input Message

HMAC Key

HMAC Engine

Usage Secret for Loaded Key 5									
0x75	0xf0	0x86	0x84	0x78	0x24	0xf8	0x79	0x39	0x5a
0x18	0x14	0x1d	0x19	0x0c	0x2f	0x01	0x29	0x0b	0x05
Output Authorization Digest									
0xef	0xb3	0xa1	0x54	0x5e	0x6c	0x48	0xe6	0xca	0xd7
0xb9	0x9e	0x4a	0x4a	0x6e	0x8d	0xb8	0x60	0xa9	0x0a

Output Digest

Figure 26.29 Calculating TPM_UnBind output authorization digest

simply a data processing machine with defined input and defined output. Any TPM vendor can guide you in making your development life easier relative to this device: for example, direct you in the procedure of turning off the penalty state so that you can freely develop applications without fear of penalty.

Most of all enjoy the experience with the knowledge that you are making your host system more secure.

Index

LIMITED WARRANTY AND DISCLAIMER OF LIABILITY

Printed and bound by CPI Group (UK) Ltd, Croydon, CR0 4YY

03/10/2024

01040336-0004